SpringerBriefs in Applied Sciences and Technology

W0079273

SpringerBriefs present concise summaries of cutting-edge research and practical applications across a wide spectrum of fields. Featuring compact volumes of 50 to 125 pages, the series covers a range of content from professional to academic.

Typical publications can be:

- A timely report of state-of-the art methods
- An introduction to or a manual for the application of mathematical or computer techniques
- A bridge between new research results, as published in journal articles
- A snapshot of a hot or emerging topic
- An in-depth case study
- A presentation of core concepts that students must understand in order to make independent contributions

SpringerBriefs are characterized by fast, global electronic dissemination, standard publishing contracts, standardized manuscript preparation and formatting guidelines, and expedited production schedules.

On the one hand, **SpringerBriefs in Applied Sciences and Technology** are devoted to the publication of fundamentals and applications within the different classical engineering disciplines as well as in interdisciplinary fields that recently emerged between these areas. On the other hand, as the boundary separating fundamental research and applied technology is more and more dissolving, this series is particularly open to trans-disciplinary topics between fundamental science and engineering.

Indexed by EI-Compendex, SCOPUS and Springerlink.

Nikolaos Kladovasilakis · Konstantinos Tsongas ·
Dimitrios Tzetzis

Innovations in Topology Optimization

With Applications in Materials, Product Design, and Manufacturing

 Springer

Nikolaos Kladovasilakis
Department of Science and Technology
Digital Manufacturing and Materials
Characterization Laboratory
International Hellenic University
Thessaloniki, Greece

Konstantinos Tsongas
Department of Industrial Engineering
and Management
School of Engineering
International Hellenic University
Thessaloniki, Greece

Dimitrios Tzetzis
Department of Science and Technology
Digital Manufacturing and Materials
Characterization Laboratory
International Hellenic University
Thessaloniki, Greece

ISSN 2191-530X   ISSN 2191-5318 (electronic)
SpringerBriefs in Applied Sciences and Technology
ISBN 978-3-031-77676-2   ISBN 978-3-031-77700-4 (eBook)
https://doi.org/10.1007/978-3-031-77700-4

This Springer imprint is published by the registered company Springer Nature Switzerland AG
The registered company address is: Gewerbestrasse 11, 6330 Cham, Switzerland

If disposing of this product, please recycle the paper.

Preface

In recent years, the field of topology optimization has undergone significant advancements, driven by the rapid development of computational design tools and the rise of additive manufacturing technologies. As these innovations continue to evolve, topology optimization has shifted from being a niche academic topic to a practical tool widely employed in both research and industry. The ability to create designs that maximize functionality, minimize material usage, and optimize mechanical performance has opened new doors for engineers and designers alike. The motivation behind writing this book stems from the increasing demand for a structured, accessible, and up-to-date resource on topology optimization. While many scientific articles and technical papers explore various aspects of this subject, they often assume prior knowledge or focus on highly specific methodologies. Our goal with this book is to bridge that gap by providing a well-rounded guide that covers the fundamental principles, common approaches, and practical applications of topology optimization in a clear and coherent manner. This book is intended to serve both students and professionals who are keen to explore the potential of topology optimization in their fields. By offering both theoretical background and hands-on examples, the content is designed to provide a balance between understanding the concepts and applying them in real-world scenarios.

The structure of the book reflects the comprehensive scope of topology optimization. The first part emphasizes on the theoretical foundation, outlining the principles and processes that underpin the methodology. Special emphasis is placed on the two most common approaches: the element-based method and the discrete or truss-based method. These sections are meant to provide readers with a solid understanding of the different techniques available and how they can be tailored to suit specific design challenges. Following the theoretical framework, the book explores practical methodologies and evaluation techniques, equipping readers with the tools to assess the efficacy of their optimized designs. A detailed discussion on computational strategies, including evaluation metrics, is included to ensure that readers can confidently apply these methods to their own work. Finally, real-world case studies, particularly from mechanical and biomechanical applications, illustrate how topology optimization is transforming industries by improving functionality, enhancing performance, and reducing material usage. These examples demonstrate the versatility and wide-ranging impact of topology optimization across different sectors. It is our hope that this book will contribute to the broader dissemination of knowledge in topology optimization and inspire further research and innovation in this fascinating field. May it serve as both a reference and a source of inspiration for those who are eager to push the boundaries of engineering design and contribute to the advancement of modern technology. Thank you for embarking on this journey with this book.

Thessaloniki, Greece   Nikolaos Kladovasilakis
                                                                          Konstantinos Tsongas
                                                                             Dimitrios Tzetzis

Acknowledgments

Many years of curiosity about topology optimization, architected materials, and understanding how the world functions have influenced and integrated these concepts into real-life applications, which are presented in this book. However, our primary focus has been on the research conducted and the discussions held over the last five years.

Hence, we would like to express our gratitude to our current place of innovation and the organization that has generously supported our research, the International Hellenic University. Without its support, the conducted research and therefore this book would not have been possible. We are particularly indebted to our talented colleagues who have provided invaluable assistance and contributions throughout our research endeavors.

We are also deeply grateful to our production editor, who not only enhanced the clarity of our writing but also challenged us to refine and elevate the key messages of the book. Additionally, we would like to express our sincere gratitude to Springer Nature for providing us the opportunity to publish this scientific book in the field of topology optimization and for their continued support in disseminating knowledge within the academic and professional community.

Finally, we extend our deepest appreciation to our families for their unwavering support, not only throughout the writing of this book but also through the many challenges that accompany an academic career. Their encouragement and understanding have been invaluable. This book is lovingly dedicated to our children, who continue to inspire us each day.

Contents

Abbreviations and Terminology

2D	Two-Dimensional
3D	Three-Dimensional
4D	Four-Dimensional
AAAM	Association for the Advancement of Automotive Medicine
ADL	Activities of Daily Living
AIS	Abbreviated Injury Scale
AM	Additive Manufacturing
ASTM	American Society for Testing and Materials
BCC	Body-Centered Cubic
BESO	Bidirectional Evolutionary Structural Optimization
CAD	Computer-Aided Design
CT	Computed Tomography
DICOM	Digital Imaging and Communications in Medicine
DM	Direct Manufacturing
EDX	Energy-Dispersive X-ray spectroscopy
ESO	Evolutionary Structural Optimization
FCC	Face Centered Cubic
FE	Finite Element
FEA	Finite Element Analysis
FEM	Finite Element Model
FOS	Factor of Safety
HA	Hydroxyapatite
HIC	Head Injury Criterion
IPC	Interpenetrating Phase Composite
ISAK	International Society for the Advancement of Kinanthropometry
MRI	Magnetic Resonance Imaging
PA	Polyamide
PBC	Periodic Boundary Conditions
PCL	Poly-Caprolactone
PLA	Poly-Lactic Acid
PSD	Particle Size Distribution

RD	Rhombic Dodecahedron
RP	Rapid Prototyping
RT	Rapid Tooling
RVE	Representative Volume Element
SD	Schwarz Diamond
SEA	Specific Energy Absorption
SEM	Scanning Electron Microscope
SIMP	Solid Isotropic Material with Penalization
SP	Schwarz Primitive
TCL	Tri-Calcium Phosphate
TO	Topology Optimization
TPMS	Triply Periodic Minimal Surfaces
TS	Thickness of Struts
WP	Weaire-Phelan

List of Figures

List of Tables

Chapter 1
Introduction

Structural optimization is the latest step in the evolution of mechanical design. The evolutionary steps of design began a few decades ago with the transition from hand-made 2D designs on paper to 2D computer-aided design (CAD) software, such as AutoCAD™. Then, the next step was the development of 3D CAD software followed by the introduction of parametric 3D CAD software, such as SolidWorks™, Inventor, NX, Fusion 360, etc., which gave designers multiple degrees of freedom to create and modify complex structures and geometries. The most recent step of this evolution is the design of structurally optimized parametric 3D CAD parts. In the last decade, structural optimization has gained increased scientific interest due to the rise of computational power and the development of advanced design software, such as nTopology™, ANSYS, etc.

Structural optimization is the process of the design of a physical structure, in order to minimize its weight or cost while meeting certain strength and stability requirements [1]. The goal of structural optimization is to create designs that are both efficient and cost-effective, while also ensuring the safety and functionality of the structure. Structural optimization of a 3D design for a component utilizes genetic algorithms and mathematical optimization techniques to optimize some specific characteristics of a part. The model consists of a set of design variables, constraints, and objective functions [2, 3]. Design variables (x) are the parameters that can be adjusted in the design of the structure or component. For example, these may include the dimensions, shape, geometry, or material properties of the structure. Constraints (g) are the limitations on the design variables and must be taken into account for a structure or component to be safe, structurally stable, and functional. These may include strength and stability requirements, material properties, and manufacturing and functionality constraints. Objective functions (f) are mathematical expressions that represent the evaluation of the design's performance. It can be the weight of the structure, the cost of the material, or other more complex performance criteria, such as the stiffness-to-weight ratio. The most common and widespread objective of the structural optimization process is to minimize the mass of material within the part's

© The Author(s), under exclusive license to Springer Nature Switzerland AG 2025
N. Kladovasilakis et al., *Innovations in Topology Optimization*,
SpringerBriefs in Applied Sciences and Technology,
https://doi.org/10.1007/978-3-031-77700-4_1

Structural Optimization

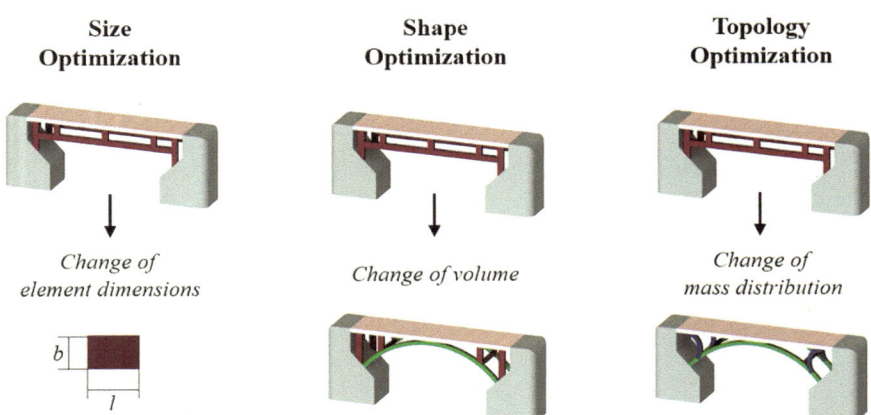

| Size
Optimization | Shape
Optimization | Topology
Optimization |

*Change of
element dimensions*   *Change of volume*   *Change of
mass distribution*

Fig. 1.1 Indicative graphical diagram with examples for the three types of structural optimization

volume domain in static loads by applying Hooke's law. The aforementioned optimization problem can be formulated as a mathematical programming problem, where the objective function is to be minimized or maximized subject to the constraints. The model of structural optimization is iterative, meaning that the design is modified and re-analyzed multiple times until an optimal solution is found and can be solved using numerical methods, such as finite element analysis (FEA), to analyze the behavior of the structure under different loads and boundary conditions. The following equation system mathematically illustrates the above-mentioned structural optimization problem, where $K(x)$ is the stiffness matrix, u is the displacement vector, and F is the applied force vector.

$$\begin{cases} \text{Minimize,} \ f(x) \\ \qquad\qquad g(x) \\ \text{s.t.,} \\ \qquad x_{\min} \leq x \leq x_{\max} \\ \text{Objective,} \ F(x) = K(x)u \end{cases} \qquad (1.1)$$

The structural optimization process can be performed utilizing mainly three methods, namely size optimization, shape optimization, and topology optimization (TO) [3]. Figure 1.1 shows the three types of structural optimization in an indicative graphical diagram applied to a bridge example.

Size optimization is a type of structural optimization that focuses on finding the optimal size of a structure or component while meeting certain strength and stability criteria. Hence, during this process, the crucial dimensions of a structure, such as the length, width, height, diameter, etc., are adjusted. It is worth noting that this process can be employed in individual components, i.e. a shaft or a beam of a structure. Size optimization can be performed using various methods, such as deterministic methods, which rely on analytical mathematical calculations or finite element methods. It is considered the simplest and the fastest type of structural optimization, due to the fact that it has one design variable (dimension) and its only constraint is to withstand the applied loading conditions.

Shape optimization involves manipulating the shape of the part to minimize weight or maximize functionality, while still meeting design requirements. This is possible by changing the size, shape, or position of the part's features. Shape optimization for functionality is a type of structural optimization that focuses on finding the optimal shape of a structure or component to enhance its performance, efficiency, and functionality. In this aspect of shape optimization the form of the structure, such as the curvature or contours of the surface, is adjusted. A characteristic example of this process is the design of parts to increase aerodynamic performance. Moreover, shape optimization can be performed using various methods, such as numerical methods (i.e. FEA) or employing generic and machine learning algorithms to explore a large number of possible shapes and identify the optimum one regarding the desired performance criteria. Shape optimization is a complex process and is mainly used in advanced engineering applications, such as aerodynamics, turbine design, biomechanics, etc., due to the fact that helps in the improvement of the performance and functionality of a component.

Topology optimization is the process that evaluates the minimum material mass and its optimal distribution within a given design volume. The procedure typically begins with a solid block of material and removes material from regions, where it is not required until it achieves the desired goal while meeting all the design requirements and constraints. The advantages of this type of structural optimization are the highly efficient, lightweight structure, increased functionality, and enhanced aesthetics. TO can be performed using a plethora of methods, which are commonly divided into two different approaches: the element-based and the discrete/truss-based approach. Each of these approaches possesses advantages and disadvantages that will be analyzed and discussed in the context of this book. TO process is the most complex type of structural optimization and it is usually performed by combining sophisticated algorithms with FEA in order to achieve the optimal result/design resulting in final designs with high geometric complexity.

Thus, the TO is superior compared to the other types of structural optimization and requires in-depth research in order to be industrialized and commercialized. In this context, the current book aims to establish a roadmap regarding the utilization of TO processes and the implementation of advanced TO techniques, such as the integration of architected materials and lattice structures. In addition, the novelty of this book is in the comprehensive presentation of architected materials and their peculiarities accompanied by practical case studies. The book's structure is organized into six chapters, with detailed descriptions for all aspects of TO. More specifically, In this chapter includes the introductory chapter, then the book consists of the presentation of the element-based TO approach (Chap. 2). Furthermore, Chap. 3 describes and summarizes the discrete/truss-based approach of TO along with a comprehensive

Fig. 1.2 Flowchart of this book

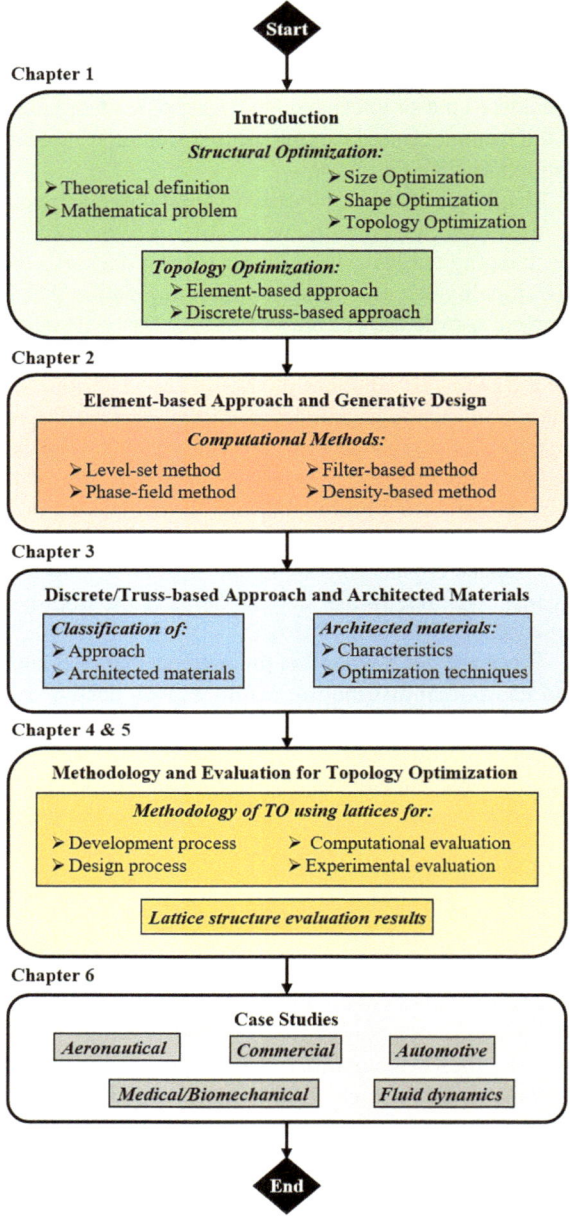

discussion of architected materials. Moreover, in Chaps. 4 and 5, the existing design, development, and evaluation methodology, focused on the usage of lattice structures, are stated coupled with the indicative evaluation results for architected materials. Finally, in Chap. 6, a series of actual case studies for topologically optimized parts is presented concerning both element-based and discrete/truss-based approaches. Figure 1.2 depicts the flowchart of this book.

References

1. U. Kirsch, *Structural Optimization* (Springer Berlin Heidelberg, Berlin, Heidelberg, 1993). https://doi.org/10.1007/978-3-642-84845-2
2. M. Ehrgott, *Multicriteria Optimization* (Springer, Berlin/Heidelberg, 2005). https://doi.org/10.1007/3-540-27659-9
3. M.P. Bendsøe, O. Sigmund, *Topology Optimization* (Springer Berlin Heidelberg, Berlin, Heidelberg, 2004). https://doi.org/10.1007/978-3-662-05086-6

Chapter 2
Element-Based Approach and Generative Design

Topology optimization has monopolized the scientific and commercial community due to the comprehensive advantages it brings to part design. Thus, the development of many 3D design software programs focused on TO, and the rapid evolution of advanced manufacturing techniques, such as additive manufacturing (AM), facilitated the implementation of TO in industrial and commercial parts [1] TO is a process that indicates the minimum mass and its optimal distribution in a given volume domain in order to withstand the applied loads, by iteratively removing or moving material until the optimal structural configuration is achieved. TO can be implemented in numerous applications in the biomechanical, aeronautical, and automotive industries [2–4]. There are two major approaches of TO, namely the element-based and discrete/truss-based, which are analyzed in the next subchapters. The element-based approach for topology optimization is a method of designing structures by iteratively removing and adding finite elements until the optimal structural design is achieved. This approach is usually used in conjunction with a stress or strain-based objective function to ensure that the final design fulfills certain structural criteria. Nowadays, there are mainly four techniques to apply the element-based TO approach, namely the level-set, phase-field, filter-based, and density-based. In the following paragraphs, each technique is described.

2.1 Level-Set Method

Level-set method is a technique of designing structures utilizing mathematical functions called a level-set function [5]. The level-set function is used to represent the optimized boundary of the structure consisting of complex shapes and topologies. This optimization process involves iteratively adjusting the level-set function to find the optimal structural configuration that meets the designer's criteria. The level-set functions are typically described as scalar functions, defined over a regular grid (finite element mesh). The level-set approach is a useful tool for topology optimization since it allows the optimum design of boundaries and could perform topological changes during the optimization process (such as adding or removing holes in the structure). However, it requires more computational resources than other approaches, and it is also more sensitive to the initialization of the level-set function.

2.2 Phase-Field Methods

The phase-field method is similar to the level-set, but it uses a smooth transition of the phase-field function between the solid and void regions, which enables a smoother transition of the material distribution [6]. More specifically, the phase-field method of TO is a technique for designing structures by representing the design domain of finite elements as a continuous field of phase variables. This method uses a mathematical function called a phase-field function that describes the distribution of materials within the volume domain. The optimization process involves iteratively adjusting the phase-field function to find the optimal structural configuration. The advantage of this process is the production of more realistic and visually appealing designs. However, the phase-field method, like the level-set method, demands more computational resources than other approaches, and it may also be more prone to be affected by the initialization of the phase-field function. Furthermore, the implementation of the method can be more challenging since it requires solving complex partial differential equations.

2.2.1 Filter-Based Method

The filter-based method of TO is a methodology for designing structures by applying a filter to the volume domain consisting of finite elements to smooth out the material distribution and eliminate small and unimportant details [7]. This method is usually employed in conjunction with other TO methods. The filter-based method can be implemented in different ways, one of the most common is the sensitivity filter, which is based on the sensitivity information of the design variables. The filter-based approach is relatively simple to implement, but it can be computationally inefficient if the filter is applied multiple times during the optimization process. Moreover, it can lead to a loss of resolution and accuracy in the final design, especially for fine details.

2.2.2 *Density-Based Method*

The density-based method of TO is a methodology for designing structures by iteratively adjusting the existence of voxels of material in the design domain until the optimal structural configuration is achieved [8]. It is the most common and widespread TO method. The optimization process starts with an initial volume domain that is divided into a large number of small finite elements (voxels). In each element, a density value is assigned between 0 and 1, with 0 representing the complete removal of the element and 1 representing full material density. The value of the density is regulated by the contribution of each voxel to the total structural performance of the designed structure, which derives from the consecutive FEAs. The elements with a crucial role in the structural integrity of the component have values of 1, in contrast to elements with a negligible contribution which have values of 0. After the execution of multiple FEAs, a final design domain is achieved consisting only of the necessary elements in order to fulfill the designer's performance criteria, such as desired strength, and maximum displacement. The density-based method of TO is a powerful tool that could potentially improve the performance and efficiency of a wide range of products and components. Furthermore, in the last five years, the TO process with the density-based method can be performed in the majority of commercial 3D design software programs via the Solid Isotropic Material with Penalization (SIMP) algorithm without extensive experience in the field, some of these platforms are SolidWorks™, ANSYS™, nTopology™, etc. This process is performed by the designers utilizing one of the aforementioned software platforms in few simple steps, as it is illustrated in the flowchart of Fig. 2.1 [9], coupled with indicative images of a biomechanical application for a topologically optimized tibial implant.

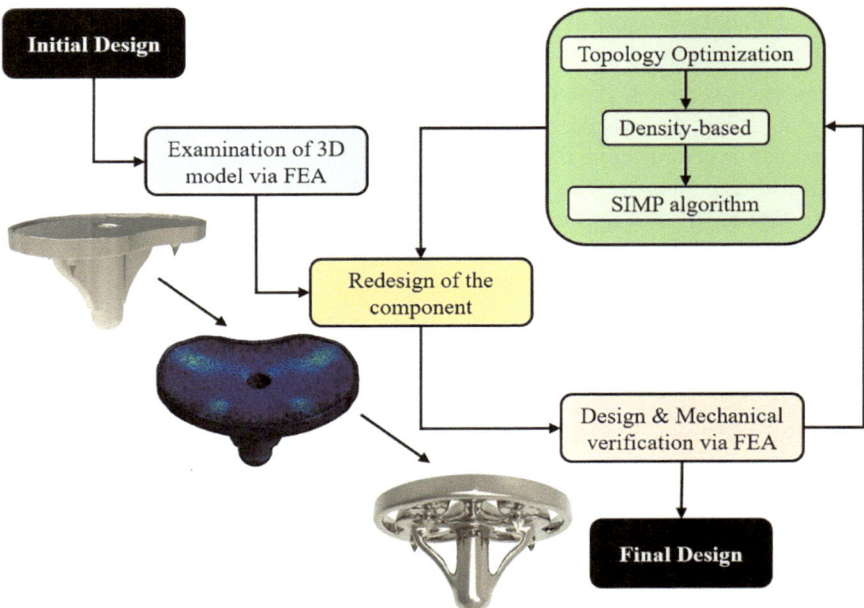

Fig. 2.1 Workflow of density-based TO process along with indicative images of a biomechanical application [9]

References

1. E. Tyflopoulos, D.T. Flem, M. Steinert, A. Olsen, State of the art of generative design and topology optimization and potential research needs (2018)
2. N. Kladovasilakis, K. Tsongas, P. Kyratsis, D. Tzetzis, Finite element analysis of hip implants with additive manufactured lattice internal geometry. Int. J. Modern Manufact. Technol. **12**(3) (2020)
3. A.W. Gebisa, H.G. Lemu, A case study on topology optimized design for additive manufacturing. IOP Conf. Ser. Mater Sci. Eng. **276**, 012026 (2017). https://doi.org/10.1088/1757-899X/276/1/012026
4. A.P. Hardwick, T. Outteridge, Vehicle light weighting through the use of Mo-lybdenum-bearing advanced high-strength steels (AHSS). Int. J. Life Cycle Assess. **21**(11), 1616–1623 (2016). https://doi.org/10.1007/s11367-015-0967-7
5. N.P. van Dijk, K. Maute, M. Langelaar, F. van Keulen, Level-set methods for structural topology optimization: a review. Struct. Multidiscip. Optim. **48**(3), 437–472 (2013). https://doi.org/10.1007/s00158-013-0912-y
6. L. Blank, H. Garcke, L. Sarbu, T. Srisupattarawanit, V. Styles, A. Voigt, *Phase-Field Approaches to Structural Topology Optimization* (2012), pp. 245–256. https://doi.org/10.1007/978-3-0348-0133-1_13
7. B. Bourdin, Filters in topology optimization. Int. J. Numer. Methods Eng. **50**(9), 2143–2158 (2001). https://doi.org/10.1002/nme.116
8. D.W. Rosen, Research supporting principles for design for additive manufacturing. Virtual Phys. Prototyp. **9**(4), 225–232 (2014). https://doi.org/10.1080/17452759.2014.951530

9. N. Kladovasilakis, T. Bountourelis, K. Tsongas, D. Tzetzis, Computational investigation of a tibial implant using topology optimization and finite element analysis. Technologies **11**, 58 (2023). https://doi.org/10.3390/technologies11020058

Chapter 3
Discrete/Truss-Based Approach and Architected Materials

3.1 Discrete/Truss-Based Approach

A discrete/truss-based approach for TO is an optimization method that focuses on the optimization of truss and lattice structures, which are usually composed of individual structural elements, such as beams and nodes, plates, or continuous surfaces [1, 2]. In the beginning, this approach was used to optimize the topology of truss-like structures, such as bridges, towers, and aircraft structures. More specifically, in this approach, the design space is divided into a grid of finite elements, and each element is assigned a binary value of 0 or 1, representing the presence or absence of a truss element at that location. Then, the optimization algorithm iteratively adjusts the binary values of the elements to find the optimal distribution of the truss elements in order to satisfy the design requirements. The objective function is typically the material mass or the cost of the truss structure, and the constraints include strength and stability requirements, such as maximum allowable stress in the truss element and minimum factor of safety against fracture, buckling, etc. One of the advantages of this approach is that it is computationally efficient, and it can handle large-scale problems with a large number of design variables. Additionally, the resulting design is often easy to interpret and understand, since it consists of a discrete set of truss elements. However, the discrete nature of this approach is also a limitation, as it results in designs that are not smooth or continuous, and it may not be able to capture some of the more complex behavior of the structure.

In recent decades, the discrete/truss-based method has been proposed with the utilization of architected materials for topology optimization [3, 4]. The design volume domain is filled with a preselected architected material in the form of a lattice structure. The optimization process involves iteratively adjusting the relative density of the lattice structure by regulating the design-related parameters, namely the strut/wall thickness and the length of the structure's unit cell. In this method, the overall mechanical response of a component can be topologically controlled. Furthermore, architected materials offer designers and engineers multiple degrees of

N. Kladovasilakis et al., *Innovations in Topology Optimization*,
SpringerBriefs in Applied Sciences and Technology,
https://doi.org/10.1007/978-3-031-77700-4_3

freedom, providing a higher level of control over material distribution by adjusting characteristics such as strength, stiffness, ductility, and relative density. The discrete/truss-based method utilizing architected materials is a particularly useful technique for topology optimization of lightweight structures, such as those used in aerospace and robotics applications. In addition, it can impart advanced physical properties in the component [5, 6] such as high porosity and high surface area to volume ratio, which are beneficial high-technology fields, including aerospace, and bioengineering. However, the main drawback of this method is that it produces structures with high geometric complexity, which can be only fabricated using advanced manufacturing systems (such as Additive Manufacturing-AM technologies). Also, this TO technique requires extensive research and development into architected materials and their physical and mechanical behavior before widespread industrial and commercial use [7]. For these reasons, the discrete/truss-based TO method utilizing architected material was chosen to be studied in this book to assist its integration into the commercial and industrial sectors. Figure 3.1 illustrates indicative images of the discrete/truss-based TO process.

The discrete/truss-based TO approach employing architected materials is a process with high complexity and requires in-depth comprehension of the nature of the architected materials, their unique topological characteristics, and the mechanisms, that influence their mechanical performance, which are also influenced by the

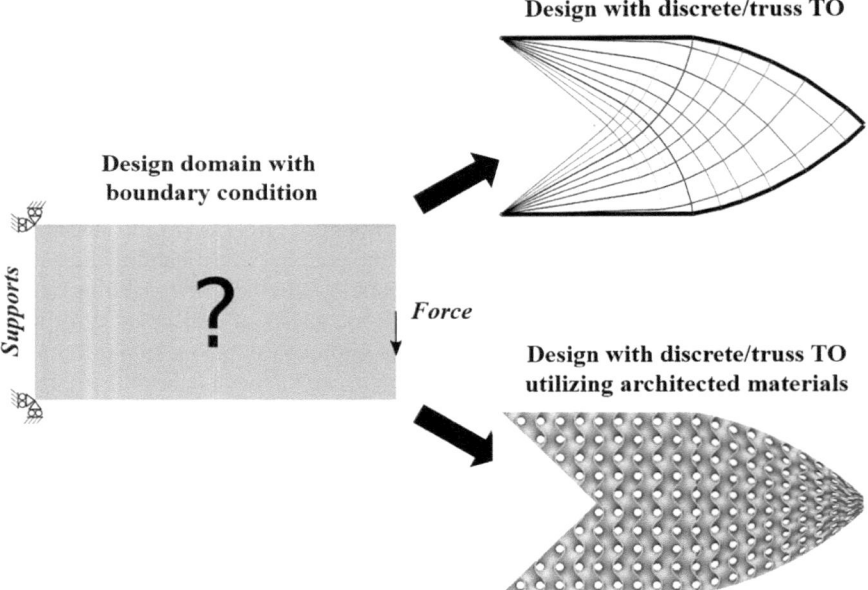

Fig. 3.1 Indicative images of discrete/truss-based TO and discrete/truss-based utilizing architected materials

design-related parameters. Thus, in the current chapter, the classification of architected materials based on geometry is analyzed in order to provide basic knowledge about the topology of existing architected materials. The theoretical background of architected material is discussed presenting the exact mathematical formulation between the mechanical properties and the design-related parameters with the relative density as an intermediate. In addition, the five most common optimization procedures for architected materials are described.

3.1.1 Classification of Architected Materials Based on the Geometry

Architected materials, also referred to as architectured materials, are a category of engineered materials that possess unique topologies and geometries, resulting in modified physical and mechanical properties of the overall structure [8]. The initial observation of such materials occurred in natural systems and in structures, such as foams, bones, and corals [9]. These natural systems possess unique characteristics, such as lightweight structures and topologically controlled mechanical behavior. As a result, the imitation of these natural cellular materials was pursued through the replication of observed natural structures or the development of new artificial architected materials, known as lattice structures. These led to a plethora of architected materials, characterized by varying geometries and topologies, necessitating their classification into categories based on their basic topological characteristics, as depicted in Fig. 3.2 [10]. The first criterion for classification is the periodicity of the structure, resulting in three wide categories of structures: stochastic, periodic, and pseudo-periodic. All existing architected materials are included in one of these three categories [11].

Stochastic architected materials lack repeating elementary geometry, meaning their topology and geometry are derived from random functions and equations [12–14]. These materials are found in nature and are not characterized by a unit cell, while artificially created stochastic architected materials are designed with the assistance and combination of random functions and topology algorithms, such as Voronoi, Delaunay, etc., to mimic natural cellular materials. Furthermore, stochastic architected materials are divided into open and closed cells, as illustrated in Fig. 3.2 [15]. The closed cells are basically composed of multiple enclosed volumes of air within the overall structure's volume. It is worth noting that AM technology allows for the precise manufacturing and examination of artificially created stochastic lattice structures with low relative densities [16].

Periodic architected materials are the majority of fabricated and examined artificial cellular materials. This is due to their ease of design, as the geometry of one unit can be repeated in three dimensions, and the predictability of their mechanical properties, such as the symmetry of the structure, can be utilized via periodic boundary conditions (PBC) to simulate their mechanical response [17]. As shown in Fig. 3.2, periodic cellular materials can be 2.5D or 3D structures [18]. 2.5D structures are the simplest

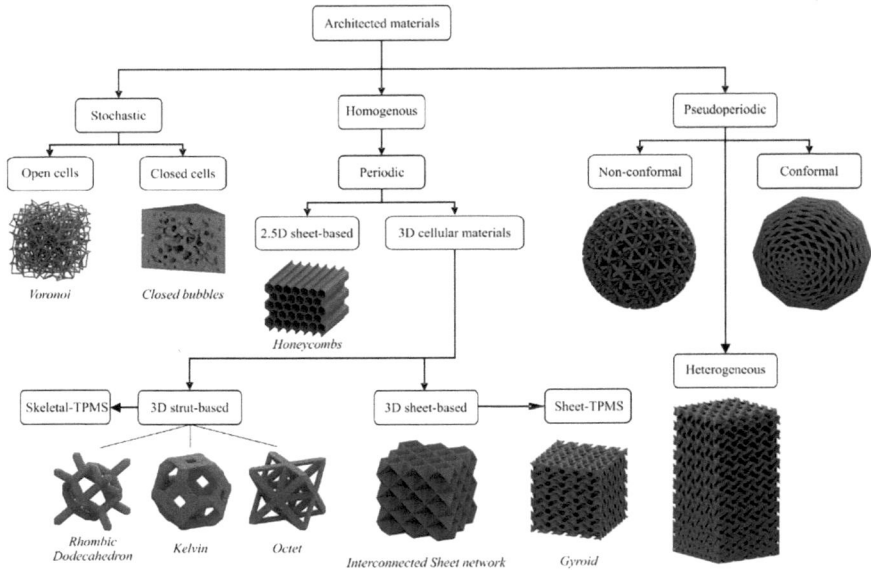

Fig. 3.2 Classification of architected materials based on the geometry [21]

form of periodic architected materials and consist of sheet networks, which are designed from 2D geometrical shapes that are extruded in the third dimension, such as honeycombs and prismatic. 3D periodic architected materials are either strut or sheet interconnected networks/surfaces. There is a wide range of strut lattices, some of the most common are the Octet, Kelvin, and Rhombic Dodecahedron, illustrated in Fig. 3.2. Sheet cellular materials are separated into shell lattices and triply periodic minimal surfaces (TPMS). Shell lattices are achieved when plates or surfaces are positioned in certain positions, usually derived from modifying strut 3D structures, as shown in Fig. 3.2. On the other hand, TPMS structures follow specific trigonometric equations that control their geometry and have the unique topology characteristic of a mean curvature equal to zero at every point of the surface [19]. Consequently, TPMS structures have a higher surface area to volume than others and are also self-supported structures, allowing for their fabrication through AM processes.

Finally, pseudo-periodic architected materials are either 2.5D or 3D cellular materials that feature a variable relative density or an interaction with the boundaries of the structure. Regarding the former, the variation in relative density occurs through changes in the periodicity of the structures, such as changes in the length of unit cells or changes in the strut/wall structure. These architected materials are also referred to as heterogeneous lattice structures. It is noteworthy that when the relative density of a structure increases gradually to distribute loads uniformly, the structure is referred to as a functionally graded lattice structure. The latter case occurs when the lattice structure conforms to the overall structure's boundary, referred to as conformal, or when the lattices are interrupted by the overall structure's boundary, referred to as non-conformal [20].

3.1.2 Theoretical Background of Architected Materials

The important physical property of the architected materials is their relative density. The relative density of an architected material is defined as the ratio of the lattice volume (V_{lattice}) to the volume of the bounding box (V_{bb}) that encapsulates the entire structure, as described by the following equation. In this equation, ρ_{solid} and m_{solid} represent the density and mass of a solid object with a volume equivalent to the bounding box volume, while ρ_{material} and m_{lattice} represent the density and mass of the lattice structure, respectively [21].

$$\overline{\rho} = \frac{\rho_{\text{lattice}}}{\rho_{\text{solid}}} = \frac{\frac{m_{\text{lattice}}}{V_{bb}}}{\frac{m_{\text{solid}}}{V_{bb}}} = \frac{m_{\text{lattice}}}{m_{\text{solid}}} = \frac{\rho_{\text{material}} \cdot V_{\text{lattice}}}{\rho_{\text{material}} \cdot V_{bb}} \rightarrow \overline{\rho(l,t)}$$
$$= \frac{V(l,t)_{\text{lattice}}}{V(l)_{bb}}. \tag{3.1}$$

Hence, the architected materials are characterized based on their applied relative density. More specifically, when the structures have an ultra-low relative density (< 5%), they are classified as foams and exhibit hyper-elastic nonlinear behavior with limited structural integrity, regardless of the construction material. These materials are suitable for impact and high-strain applications. Conversely, when the relative density is within the range of 10–50%, the architected materials are referred to as lattice structures. Depending on the lattice geometry and the applied relative density, they can be utilized for structural or energy absorption purposes. It should be noted that architected materials with relative densities above 60% possess mechanical behavior close to that of solid objects [22]. Moreover, the gray area of relative density between each class occurs due to the fact that the architected material could behave as one of the two classes based on its topology, construction material, and manufacturing technique.

The relative density of periodic architected materials is closely associated with the ratio between the thickness (t) of the strut (for strut lattices) or the wall (for sheet/surface) to the length (l) of the structure's unit cell. In contrast, in stochastic structures, the relative density is dependent on the ratio of the strut/wall thickness to the distance between the random seeds of the stochastic structure. According to existing literature [21, 23, 24], these ratios influence the relative densities of architected materials in different ways, depending on the geometry and topology of the structures. As an example, for the majority of sheet-TPMS lattices. the t/l ratio has a linear relation with the relative density, as described by Eq. (4.2). On the contrary, the majority of 2.5D lattices (honeycombs, etc.) and 3D strut lattices (octet, diamond) influence the relative density with a second- or third-order polynomial mathematical formulation, depending on the structure's geometry and the accuracy of the calculations, as outlined in Eqs. (4.3) and (4.4). It is important to mention that the constants C_1, C_2, and C_3 are calculated based on the employed architected material. Additionally, t is always lower than l, frequently $t \ll l$, resulting in $t/l < 1$. Therefore, in strut lattices, the relative density increases exponentially when t/l is low, in contrast with sheet-TPMS lattices, where the relative density increases linearly. It is worth mentioning that the following equations calculate the relative density with a high accuracy of $\pm\,0.25\%$ and are applicable for relative density up to 50% due to the overlapping of the structure's elements.

$$\overline{\rho} = C_1 \cdot \frac{t}{l}. \tag{3.2}$$

$$\overline{\rho} = C_2 \cdot \left(\frac{t}{l}\right)^2. \tag{3.3}$$

$$\overline{\rho} = C_2 \cdot \left(\frac{t}{l}\right)^2 - C_3 \cdot \left(\frac{t}{l}\right)^3. \tag{3.4}$$

3.2 Optimization Methods for Architected Materials

Architected materials are utilized to improve the structural design of a part by reducing its weight and distributing mass more evenly. However, there are methods that assist in the further optimization of these materials by exploiting their unique topological characteristics. According to the literature [18, 25–28], there are five widespread methods for optimizing architected materials and lattice structures, which is a research field that has drawn more attention from the scientific community during the last decade.

The first and most common method for optimizing architected materials and lattice structures is functional gradation. This technique is a second-level topology optimization method and as in the first level, the replacement of solid region by the selected lattice structure is performed. Furthermore, this method involves the adjustment of the local relative density within the design domain based on the applied loads. As an example, in regions with low-stress concentrations, lower relative density is applied compared to the regions with higher stress concentrations, where the relative density is denser [15]. This improves the mechanical performance of the structure without changing its overall mass because its overall relative density remains the same. The implementation of functional gradation could vary depending on the design process, the application, etc. More specifically, the changes in relative density could be steep increasing from 10 to 60% relative density for two contiguity regions, or it could be employed gradually with a linear, exponential, or scalar-field mode based on the stress concentration contours/field providing a smoother transaction between the regions of low and high relative densities. Additionally, functional gradation of relative density can be achieved by altering the t/l ratio (by adjusting either the numerator or the denominator). Figure 3.3a shows an example of how this method is applied in an architected material (Gyroid) for impact and energy absorption application.

The second technique for optimizing architected materials involves an additional manufacturing process besides the creation of the architected material. The concept of this procedure is to fill the void inside the volume with other material, as it is depicted in Fig. 3.3b. This process creates solid composite structures consisting of two materials, i.e. the architected material and a filling material. These structures are also known as interpenetrating phase composites. There is a wide variety of filling materials from epoxy and silicone rubber compounds to polymer foams, such as polyurethane. In addition, the interpenetrating phase composites with architected materials could produce composite materials with unique physical and mechanical properties, such as structures with high electrical conductivity or with higher energy absorption rates [29, 30].

The third method employs two or more different architected materials inside the overall volume of a component. This method is also known as the hybridization of lattice structures [31]. In detail, a different architected material is utilized for a different region of the structure with the goal of improving the overall mechanical behavior, exploiting the unique physical, mechanical, or aesthetic properties of each structure [32]. For instance, auxetic architected materials can be employed in areas where tensile stresses are present, while sheet-TPMS, such as the Schwarz Diamond (SD) structure, can be utilized to secure the structural integrity of a component. Additionally, the hybridization of lattice structures can reveal remarkable results in impact applications, by maximizing the energy absorption of the entire structure through the combination of a bending-dominated architected material, which undergoes maximum plastic deformation, and a stretching-dominated lattice structure, that experiences elastic deformation with high peak strength. The process of hybridization can be implemented rapidly or gradually, as depicted in Fig. 3.3c, which illustrates a hybrid lattice designed for bending and energy absorption tests.

The fourth method of optimization involves fundamentally altering the unit cell structure of an architected material through the combination of two or more distinct architected materials within the volume of a single unit cell, thereby creating an advanced cellular material with enhanced mechanical properties [33]. Similar to the previous method, this method of hybridization is performed at the level of the unit cell. The hybridization process involves evaluating the individual mechanical performance of various architected materials and then combining them in a manner that strengthens and reinforces regions of stress concentration, resulting in a new architected material with improved mechanical behavior and more uniform stress distribution. This approach enables the creation of advanced cellular materials with high geometric complexity, which can be paired with the previously mentioned optimization methods of lattice structures for applications requiring high levels of structural integrity. Figure 3.3d illustrates unit cells of the hybrid architected materials and their initial cellular materials, along with their contours of von Mises stress, which were used to determine the appropriate combination of structures.

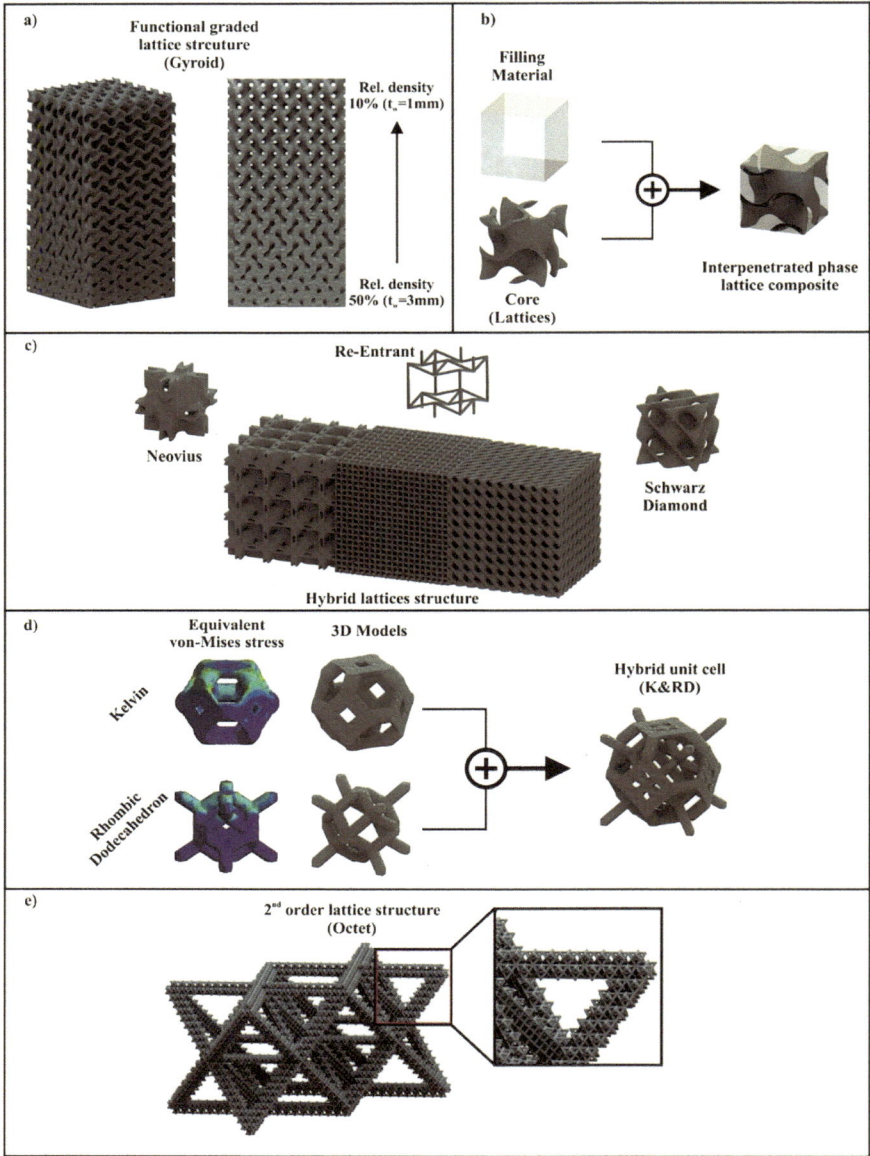

Fig. 3.3 Optimization technique of architected materials: **a** functional gradation, **b** interpenetrating phase composites, **c** hybridization of a structure, **d** hybridization of a unit cell, **e** higher-order lattice structure

The final optimization technique results in the creation of highly complex lattice structures through the production of architected materials with higher-order structures [34]. Specifically, this approach involves embedding a lattice structure within the elements, such as struts or walls and nodes, of an architected material, as depicted in Figure 3.3e. This method is employed to create ultra-lightweight and highly porous structures. However, a significant challenge of this optimization method is the manufacturing process, as the lattice structures embedded within the struts or surfaces have dimensions that are one order smaller than the overall structure. Therefore, this approach requires the employment of a 3D printer with a large build volume and high accuracy, such as the selective laser melting (SLM) AM technique, or a 3D printer with ultra-high accuracy, such as the two-photon polymerization AM technique.

References

1. M.P. Bendsøe, A. Ben-Tal, J. Zowe, Optimization methods for truss geometry and topology design. Struct. Optim. **7**(3), 141–159 (1994). https://doi.org/10.1007/BF01742459
2. W. Achtziger, On simultaneous optimization of truss geometry and topology. Struct. Multi-discip. Optim. **33**(4–5), 285–304 (2007). https://doi.org/10.1007/s00158-006-0092-0
3. C. Pan, Y. Han, J. Lu, Design and optimization of lattice structures: a review. Appl. Sci. **10**(18), 6374 (2020). https://doi.org/10.3390/app10186374
4. V.-N. Hoang, P. Tran, V.-T. Vu, H. Nguyen-Xuan, Design of lattice structures with direct multi-scale topology optimization. Compos. Struct. **252**, 112718 (2020). https://doi.org/10.1016/j.compstruct.2020.112718
5. P. Heinl, L. Müller, C. Körner, R.F. Singer, F.A. Müller, Cellular Ti–6Al–4V structures with interconnected macro porosity for bone implants fabricated by selective electron beam melting. Acta Biomater. **4**(5), 1536–1544 (2008). https://doi.org/10.1016/j.actbio.2008.03.013
6. M.R. Dias, J.M. Guedes, C.L. Flanagan, S.J. Hollister, P.R. Fernandes, Optimization of scaffold design for bone tissue engineering: a computational and experimental study. Med. Eng. Phys. **36**(4), 448–457 (2014). https://doi.org/10.1016/j.medengphy.2014.02.010
7. S.D. Larsen, O. Sigmund, J.P. Groen, Optimal truss and frame design from projected homogenization-based topology optimization. Struct. Multi-disciplinary Optim. **57**(4), 1461–1474 (2018). https://doi.org/10.1007/s00158-018-1948-9
8. T.A. Schaedler, W.B. Carter, Architected cellular materials. Annu. Rev. Mater. Res. **46**(1), 187–210 (2016). https://doi.org/10.1146/annurev-matsci-070115-031624
9. L.J. Gibson, Modelling the mechanical behavior of cellular materials. Mater. Sci. Eng. A **110**, 1–36 (1989). https://doi.org/10.1016/0921-5093(89)90154-8
10. N. Kladovasilakis, K. Tsongas, D. Karalekas, D. Tzetzis, Architected materials for additive manufacturing: a comprehensive review. Materials **15**(17), 5919 (2022). https://doi.org/10.3390/ma15175919
11. E. Pei, I. Kabir, T. Breški, D. Godec, A. Nordin, A review of geometric dimensioning and tolerancing (GD&T) of additive manufacturing and pow-der bed fusion lattices. Prog. Addit. Manufact. **7**(6), 1297–1305 (2022). https://doi.org/10.1007/s40964-022-00304-8
12. S. Kanwar, O. Al-Ketan, S. Vijayavenkataraman, A novel method to design biomimetic, 3D printable stochastic scaffolds with controlled porosity for bone tissue engineering. Mater. Des. **220**, 110857 (2022). https://doi.org/10.1016/j.matdes.2022.110857
13. J.-H. Groth, C. Anderson, M. Magnini, C. Tuck, A. Clare, Five simple tools for stochastic lattice creation. Addit. Manuf. **49**, 102488 (2022). https://doi.org/10.1016/j.addma.2021.102488
14. M.J. Mirzaali, R. Hedayati, P. Vena, L. Vergani, M. Strano, A.A. Zadpoor, Rational design of soft mechanical metamaterials: independent tailoring of elastic properties with randomness. Appl. Phys. Lett. **111**(5), 051903 (2017). https://doi.org/10.1063/1.4989441

15. D. Mahmoud, M. Elbestawi, Lattice structures and functionally graded materials applications in additive manufacturing of orthopedic implants: a review. J. Manufact. Mater. Process. **1**(2), 13 (2017). https://doi.org/10.3390/jmmp1020013
16. I. Gibson, D.W. Rosen, B. Stucker, *Additive Manufacturing Technologies* (Springer US, Boston, MA, 2010). https://doi.org/10.1007/978-1-4419-1120-9
17. F. Derveni, A.J. Gross, K.D. Peterman, S. Gerasimidis, Postbuckling behavior and imperfection sensitivity of elastic-plastic periodic plate-lattice materials. Extreme Mech. Lett. **50**, 101510 (2022). https://doi.org/10.1016/j.eml.2021.101510
18. J. Bauer, L.R. Meza, T.A. Schaedler, R. Schwaiger, X. Zheng, L. Valdevit, Nanolattices: an emerging class of mechanical metamaterials. Adv. Mater. **29**(40), 1701850 (2017). https://doi.org/10.1002/adma.201701850
19. Z. Dong, X. Zhao, Application of TPMS structure in bone regeneration. Eng. Regeneration **2**, 154–162 (2021). https://doi.org/10.1016/j.engreg.2021.09.004
20. J. Corona-Castuera, D. Rodriguez-Delgado, J. Henao, J.C. Castro-Sandoval, C.A. Poblano-Salas, Design and fabrication of a customized partial hip prosthesis employing CT-scan data and lattice porous structures. ACS Omega **6**(10), 6902–6913 (2021). https://doi.org/10.1021/acsomega.0c06144
21. N. Kladovasilakis, K. Tsongas, I. Kostavelis, D. Tzovaras, D. Tzetzis, Effective mechanical properties of additive manufactured triply periodic minimal surfaces: experimental and finite element study. Int. J. Adv. Manufact. Technol. **121**(11–12), 7169–7189 (2022). https://doi.org/10.1007/s00170-022-09651-w
22. L.J. Gibson, M.F. Ashby, *Cellular Solids* (Cambridge University Press, 1997). https://doi.org/10.1017/CBO9781139878326
23. V.S. Deshpande, N.A. Fleck, M.F. Ashby, Effective properties of the octet-truss lattice material. J. Mech. Phys. Solids **49**(8), 1747–1769 (2001). https://doi.org/10.1016/S0022-5096(01)00010-2
24. N. Kladovasilakis, K. Tsongas, I. Kostavelis, D. Tzovaras, D. Tzetzis, Effective mechanical properties of additive manufactured strut-lattice structures: experimental and finite element study. Adv. Eng. Mater. **24**(3), 2100879 (2022). https://doi.org/10.1002/adem.202100879
25. O. Al-Ketan, M. Adel Assad, R.K. Abu Al-Rub, Mechanical properties of periodic interpenetrating phase composites with novel architected microstructures. Compos. Struct. **176**, 9–19 (2017). https://doi.org/10.1016/j.compstruct.2017.05.026
26. Z. Chen, Y.M. Xie, X. Wu, Z. Wang, Q. Li, S. Zhou, On hybrid cellular materials based on triply periodic minimal surfaces with extreme mechanical properties. Mater. Des. **183**, 108109 (2019). https://doi.org/10.1016/j.matdes.2019.108109
27. N. Novak, O. Al-Ketan, M. Borovinšek, L. Krstulović-Opara, R. Rowshan, M. Vesenjak, Z. Ren, Development of novel hybrid TPMS cellular lattices and their mechanical characterisation. J. Market. Res. **15**, 1318–1329 (2021). https://doi.org/10.1016/j.jmrt.2021.08.092
28. O. Al-Ketan, R.K. Abu Al-Rub, Multifunctional mechanical metamaterials based on triply periodic minimal surface lattices. Adv. Eng. Mater. **21**(10), 1900524 (2019). https://doi.org/10.1002/adem.201900524
29. S. Berhanu, F. Tariq, T. Jones, D.W. McComb, Three-dimensionally inter-connected organic nanocomposite thin films: implications for donor-acceptor photovoltaic applications. J. Mater. Chem. **20**(37), 8005 (2010). https://doi.org/10.1039/c0jm01030h
30. Y. Liu, D.L. Duan, S.L. Jiang, S. Li, Preparation and its cavitation performance of nickel foam/epoxy/SiC co-continuous composites. Wear **332–333**, 979–987 (2015). https://doi.org/10.1016/j.wear.2014.12.025
31. N. Novak, O. Al-Ketan, M. Borovinšek, L. Krstulović-Opara, R. Rowshan, M. Vesenjak, Z. Ren, Development of novel hybrid TPMS cellular lattices and their mechanical characterisation. J. Mater. Res. Technol. **15**, 1318–1329 (2021). https://doi.org/10.1016/j.jmrt.2021.08.092
32. L. Cui, F. Jiang, D. Deng, T. Xin, X. Sun, R.T. Mousavian, R.L. Peng, Z. Yang, J. Moverare, Cyclic response of additive manufactured 316L stainless steel: the role of cell structures. Scr. Mater. **205**, 114190 (2021). https://doi.org/10.1016/j.scriptamat.2021.114190

33. Z.P. Sun, Y.B. Guo, V.P.W. Shim, Characterisation and modeling of additively-manufactured polymeric hybrid lattice structures for energy absorption. Int. J. Mech. Sci. **191**, 106101 (2021). https://doi.org/10.1016/j.ijmecsci.2020.106101
34. S. Al Hassanieh, A. Alhantoobi, K.A. Khan, M.A. Khan, Mechanical proper-ties and energy absorption characteristics of additively manufactured light-weight novel re-entrant plate-based lattice structures. Polym. (Basel) **13**(22), 3882 (2021). https://doi.org/10.3390/polym13223882

Chapter 4
Methodology for Topology Optimization

Topology optimization is a powerful computational technique used to determine the most efficient material layout within a given design space, subject to predefined constraints and loads. This method is widely employed in engineering disciplines to enhance performance, reduce weight, and improve the overall functionality of structures and components. The main topology optimization process involves iteratively refining the design to meet specified objectives, often leading to innovative and non-intuitive configurations. However, in the approach that the topology optimization is derived with the utilization of architected materials, the followed implementation methodology is more complex. More specifically, the most essential factors for the utilization of architected materials are their experimental mechanical behavior and numerical evaluation methodology, which are presented in-depth in Sect. 4.2.

4.1 Methodology for Element-Based Approach

The density-based method is the most common and widespread method of TO processes to achieve designs that require the minimum amount of material (mass) while fulfilling the desired structural criteria. The success of this method can be attributed to the employment of automatic, fully functional, and effective algorithms. The most common algorithms for density-based TO are ESO, BESO, and SIMP [1]. Evolutionary structural optimization (ESO) is based on the simple concept of gradually removing inefficient material mass from the structure's volume [2]. The resulting structure of the ESO method evolves toward its optimal shape and topology. The ESO method is quite helpful for engineers focused on investigating structurally effective shape forms during the conceptual design stage. The ESO method is also called a hard destruction technique since it enables repetitive removal or addition of finite amounts of material [3]. Heuristic criteria are employed that can be based on well-defined sensitivity information. Therefore, the ESO method is quite easy

N. Kladovasilakis et al., *Innovations in Topology Optimization*,
SpringerBriefs in Applied Sciences and Technology,
https://doi.org/10.1007/978-3-031-77700-4_4

to implement, which is an advantage for TO problems involving complex physical processes. ESO technique along with FEA estimates the stress level of an arbitrary partition of a structure [2]. A low level of stress in the individual part of a structure helps to determine the ineffective use of the material. It is preferable to distribute the stress level all over the structure according to the safety factor. Based on that, the principle of material removal has been established, where insufficiently loaded material is removed from individual elements of the FEM [4]. Comparing the stress σ_{evm} element with the critical or maximum value σ_{maxvm} of the object determines the stress level of that element. If the element satisfies the following condition:

$$\frac{\sigma_e^{vm}}{\sigma_{max0}^{vm}} < R_{Ri}, \qquad (4.1)$$

where R_R is the limit value (subtraction rate) then this element is removed and the control process starts a new control cycle. The test of the individual components is performed into iterations until a steady state is reached, i.e. the state in which there are no other components that fulfill this subtraction limit. According to the specific evolution rate H_i the subtraction rate can then be increased by the equation:

$$R_{R(i+1)} = R_{Ri} + H_i. \qquad (4.2)$$

The above process can be applied for an increased subtraction ratio and the FEAs are performed until a new steady-state equilibrium is achieved. For instance, until all the material is removed from those areas, where the stress level does not exceed 20% of the maximum material stress. A quantitative evaluation of the stiffness change of the structure because of the subtraction of i-th finite element is the elasticity index, defined for the mean elasticity as shown in the following equation, where u_i is the node displacement vector of element i and K_i is the element's stiffness matrix.

$$\alpha_i^e = \frac{1}{2} * u_i^T K_i \qquad (4.3)$$

The sensitivity function reveals the decrease of the mean stiffness, as a result of the removal of the i-th element, which is equal to the elemental deformation energy of the i-th element. To maintain the stiffness by removing the elements, it is necessary to remove elements with the minimum value of the sensitivity factor. The mathematical formulation of the ESO algorithm is equally applicable for 2D and 3D problems as it is quite simple and clear. The subtraction of elements is accomplished by assigning a zero value to their coefficient of the equation, and as a result, they are ignored during the subsequent repetitions. This iterative process of removing data leads to a reduction in the number of equations, thus the computational demands of the problem are diminished, which is particularly crucial for 3D problems. A major disadvantage of the ESO method is that it does not allow the removed material to be recovered, while this material may be efficient in subsequent ones. To summarize, it is obvious that the ESO method in some cases, does not provide the optimal solution and this disadvantage is eliminated with the BESO method.

The bidirectional evolutionary structural optimization (BESO) algorithm allows the simultaneous removal and addition of material to the design volume [1]. The fundamental difference between these two methods is that the sensitivity index of the empty elements is determined by linear extrapolation of the displacement field, obtained as a result of FEA. After that, the full elements with the lowest sensitivity index values are removed from the structure and the empty elements with the highest sensitivity values are filled with material. The number of elements removed and added at each iteration is determined by two independent parameters: the R_R subtraction ratio and the R_I inclusion ratio.

The third and most widespread method is the solid isotropic material with penalization (SIMP) algorithm. It is worth mentioning that the majority of commercial design software utilizes this algorithm of TO. The fundamental idea of the SIMP method is to generate a virtual density field that is proportional to the actual characteristics of the understudy structure. SIMP technique decreases the compliance of the structure due to the reallocation of the material [5] in the examination area under specific limit conditions. The result of using the SIMP method is that the object maintains the same stiffness values within the under-consideration area. SIMP method has been extensively employed in additive manufacturing constructs. The material density is considered a design variable for the calculation stage of the optimization. Therefore, the optimal structure, within a planned area, is achieved by redistributing the material based on the criteria given during the optimization. The SIMP method is also based on dividing the examined volume into voxels. The material properties are kept constant in each of these elements and depend on the relative density x_i. When the optimization process is terminated, the relative density of each element must be equal to one or zero. To limit the intermediate relative density, the rejection factor p is used. As the design variables are assigned, the relative densities of the elements and the mean correspondence are chosen as the objective function. The problem of topology optimization for minimum correspondence can be expressed as [5]:

$$\text{Find} : X = \{x_1, x_2, \ldots, x_i\}^T, i = 1, 2, \ldots, n$$
$$\downarrow$$
$$\text{Min} : C(X) = F^T U = U^T K U = \sum_{i=1}^{n} u_i^T k_i u_i = \sum_{i=1}^{n} (x_i)^P u_i^T k_0 u_i$$
$$\downarrow$$
$$\text{Subj.to} : K U = F, V = f_0 V_0 = \sum_{i=1}^{n} x_i v_i \tag{4.4}$$
$$\downarrow$$
$$\text{with} : 0 < x_{\min} \leq x_i \leq x_{\max} \leq 1$$

where

- C is the objective function and is defined as the mean correspondence
- X is the vector of construction variables

- F is the loading vector
- U is the total displacement vector
- K is the total stiffness strain
- V is the material's volume
- F_0 is the volumetric ratio.

The last and most efficient density-based algorithm is the ESO-SIMP method. It is a combination of the ESO and SIMP methods and aims to compensate for the disadvantages of these two topology optimization methods [6]. To solve the optimization problem, the relative densities are used as the designed variables and the mean correspondence is selected as the objective function. The optimization problem for the minimum mn correspondence based on the ESO-SIMP algorithm is as follows:

$$\text{Find} : X = \{x_1, x_2, \ldots, x_i\}^T, i = 1, 2, \ldots, n$$
$$\downarrow$$
$$\text{Min} : C(X) = U^T K U = \sum_{i=1}^{n} u_i^T k_i u_i = \sum_{i=1}^{n} (x_i)^P u_i^T k_0 u_i$$
$$\downarrow$$
$$\text{Subj. to} : KU = F, V = \sum_{i=1}^{n} x_i v_i \le f_0 V_0$$
$$\downarrow$$
$$\text{with} : 0 < x_{\min} \le x_i \le x_{\max} \le 1$$

$$(4.5)$$

The difference between the ESO-SIMP and SIMP methods is the volume limitation. During each iteration, elements whose relative density is less or equal to the rejection factor are removed from the design volume and all the other elements are inserted into the next iteration. This combination method turns out to be more appropriate than an individual ESO or SIMP in terms of efficiency and reliability.

4.2   Methodology for TO with Architected Materials

4.2.1   Mechanical Behavior

As it was mentioned the implementation of architected materials via lattice structures for topology optimization purposes is a complex process. In the previous chapter, it was described mathematically the influence of design-related parameters on the relative density, hence, the next step is the connection of relative density with the mechanical behavior of the architected materials. According to the existing literature [7–9], the mechanical properties of an architected material are indissolubly connected with the applied relative density obeying a scaling law because of the size effect. This size effect is the mechanism that reveals how the mechanical behavior is compromised by the variation of the relative density and it mainly affects architected materials with relative densities below 50%. The following equation describes the above-mentioned scaling law, where Φ_{solid} is the mechanical property of the construction material, Φ_{lattice} is the effective mechanical property for the lattice structure, and G and n are constants that their values depend on the construction material and the applied architected material.

$$\frac{\Phi_{\text{lattice}}}{\Phi_{\text{solid}}} = G \cdot (\overline{\rho})^n \tag{4.6}$$

In Fig. 4.1, an example of the application of the effective mechanical properties on a stochastic architected material is presented. In detail, assuming that for a specific application, the stiffness (K_e) and the maximum applied force (F_{max}) must be found. According to Hooke's law, the cross-section of the specimens (A_0) would be multiple with the construction material's peak stress ($\sigma_{m,\text{peak}}$) in order to extract the $\overline{F}_{\text{max}}$ and the stiffness would be calculated, according to the following equations (Eq. 4.7), where L is the height of the specimen and E is the elastic modulus of the construction material.

$$F_{\text{max}} = \sigma_{m,\text{peak}} \cdot A_0 \quad \text{and} \quad K_e = \frac{E \cdot A_0}{L} \tag{4.7}$$

However, the above-mentioned methodology is not applicable in complex architected materials due to the multiple changes in the cross-section areas ($A_l \neq A_u$). Thus, through the characterization of the architected materials and the extraction of the effective mechanical properties, such as the effective elastic modulus (E_{eff}) and

Example of Effective Mechanical Properties

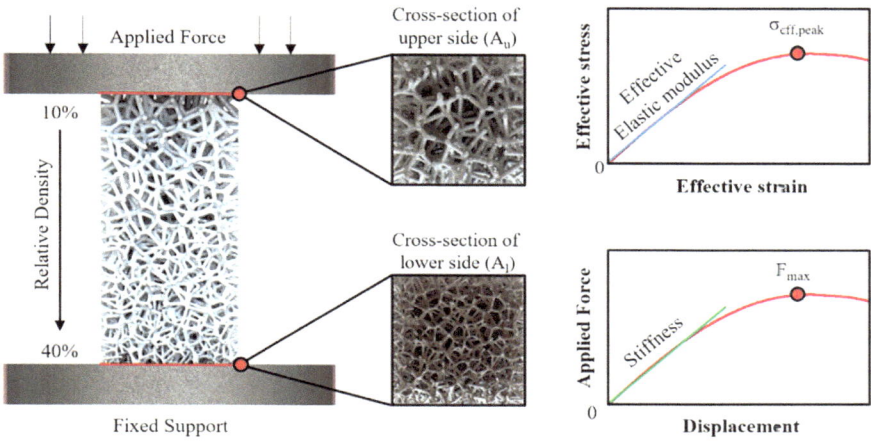

Fig. 4.1 Example of effective mechanical properties for a stochastic architected material

the effective peak stress ($\sigma_{\text{eff,peak}}$), it is possible to overcome this obstacle and calculate the structures' stiffness and maximum force by utilizing the same set of equations (Eq. 4.7) and replacing the material properties with the effective mechanical properties. This method is essential in applications where the stiffness and the strength of the lattice structure should be comparable with the properties of the construction material, such as in biomechanical applications to avoid the stress shielding effect.

Furthermore, according to Eq. (4.6), the applied relative density influences the effective mechanical properties of an architected material in a nonlinear manner. It is obvious that if the n is above one leads to exponential behavior. However, if the n is equal to or below one results in a linear or logarithmic influence, respectively. In the context of this book, the above-listed equations (Eq. 3.2), (Eq. 3.4), and (Eq. 4.6) were combined leading to a direct connection of design-related parameters with the expected effective mechanical properties of an architected material for a given relative density, as it is depicted in the following equation (Eq. 4.8). These are important tools for the employment of architected materials in potential applications [10, 11].

$$\frac{\Phi_{\text{lattice}}}{\Phi_{\text{solid}}} = G \cdot (\overline{\rho})^n \begin{cases} \frac{\Phi_{\text{lattice}}}{\Phi_{\text{solid}}} = G \cdot C_1^n \cdot \left(\frac{t}{l}\right)^n & \text{For TPMS lattices} \\ \frac{\Phi_{\text{lattice}}}{\Phi_{\text{solid}}} = G \cdot \left(C_2 \cdot \left(\frac{t}{l}\right)^2 - C_3 \cdot \left(\frac{t}{l}\right)^3\right) & \text{For strut lattices} \end{cases} \quad (4.8)$$

Research endeavors [7] so far have established that the effective mechanical properties of an architected material are dependent on the relative density of the structure, in accordance with the previously mentioned scaling law (Eq. 4.6). Through extensive research [12], two major mechanical behaviors have been identified based on the exponent of the scaling law for the effective elastic modulus of a lattice structure and the curve of effective engineering stress to strain diagrams: the stretching-dominated and the bending-dominated behaviors. It is worth noting that actual experimental data on the physically architected materials are presented in the following chapters based on the results of published studies and the results of the current book.

The stretching-dominated behavior occurs in architected materials with high connectivity, either with struts or surfaces and limited degrees of freedom in each joint/node, leading to numerous possible self-stress situations. Under loading conditions, the initial response of the structure is to elastically stretch the struts/surfaces, resulting in a linear relationship between the applied relative density and the effective elastic modulus (E_{lattice}) and the effective yield strength ($\sigma_{\text{yield,lattice}}$), as described by the following equations, and the exponents (n_E and n_σ) of the scaling laws being close to one. Furthermore, Figure 4.2a illustrates an indicative stress–strain diagram for a stretching-dominated behavior.

$$\frac{E_{\text{lattice}}}{E_{\text{solid}}} = G_{Es} \cdot \overline{\rho}^{n_E}, \quad (4.9)$$

$$\frac{\sigma_{\text{yield,lattice}}}{\sigma_{\text{yield,solid}}} = G_{\sigma,s} \cdot \overline{\rho}^{n_\sigma}, \quad (4.10)$$

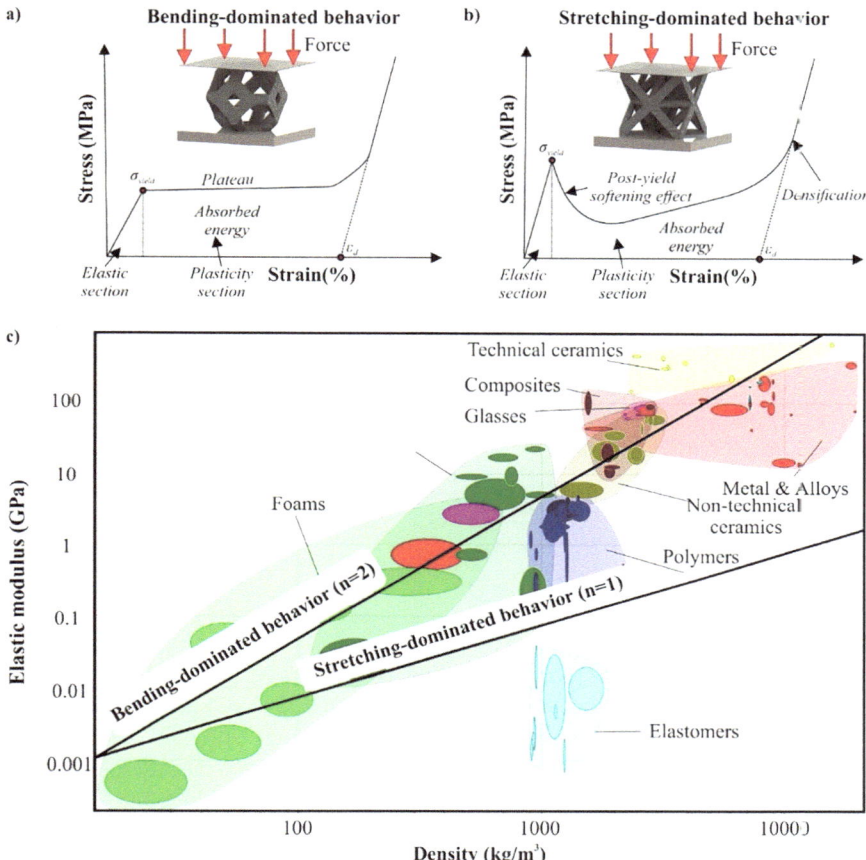

Fig. 4.2 Indicative mechanical response of an architected material for: **a** stretching-dominated behavior (Octet), **b** bending-dominated behavior (Kelvin), **c** effective elastic modulus to relative density bubble chart along with the curves for the two examined mechanical behaviors

where E_{solid} and $\sigma_{\text{yield,solid}}$ are the elastic modulus and the yield strength of the solid material, and G_{Es} and G_{σ} are constants for stretching-dominated behavior depending on the geometry and material of the applied lattices. Architected materials that exhibit stretching-dominated behavior or close to stretching-dominated behavior ($n_E \approx n_{\sigma} \approx 1$) reveal higher stiffness than lattices with bending-dominated behavior and have a higher peak strength due to their connectivity. However, the intense loading of the struts/surfaces at peak strength results in plastic buckling or brittle fracture of most of them, causing a rapid decrease in the strength of the structure after the peak. This phenomenon is known as the post-softening effect and is reflected in the stress–strain diagram as a rapid drop in the stress–strain curve after maximum stress, as shown in Fig. 4.2a. Therefore, architected materials with stretching-dominated behavior are suitable for applications that require enhanced structural integrity. Nevertheless, these structures have moderate performance in terms of energy absorption due to the post-softening effect, which results in most of the energy being absorbed in the elastic section.

On the other hand, the bending-dominated behavior is observed in architected materials with low connectivity, allowing the struts/surfaces within the structure to bend under the applied loads. As a result of the struts/surfaces bending, elastic deformation is governed by the surface's second moment of inertia, leading to an exponential relationship between the effective elastic modulus and the relative density. The mathematical expression of this relationship is presented in the following equations, where G_{Eb} and $G_{\sigma,s}$ are constants for bending-dominated behavior that are dependent on the geometry and material of the architected material. Exponent n_E has values close to two or greater ($n_E \geq 2$) and n_σ has values equal to or higher than 1.5 ($n_\sigma \geq 3/2$). Furthermore, Figure 4.2b illustrates an indicative effective stress–strain diagram for a bending-dominated behavior.

$$\frac{E_{\text{lattice}}}{E_{\text{solid}}} = G_{\text{Eb}} \cdot (\overline{\rho})^{n_E} \tag{4.11}$$

$$\frac{\sigma_{\text{yield,lattice}}}{\sigma_{\text{yield,solid}}} = G_{\sigma,s} \cdot (\overline{\rho})^{n_\sigma} \tag{4.12}$$

Architected materials that exhibit bending-dominated behavior have usually low stiffness and peak strength due to the bending of struts/surfaces and the low connectivity of the structure's elements. However, after peak stress and due to the bending loading condition of the struts/surfaces of the architected materials, these structures experience constant stresses, leading to a plateau region in the stress–strain diagrams, as depicted in Fig. 4.2. During the plateau, the structure undergoes plastic deformation until the densification section (after 50% of strain). This plastic deformation leads to significant energy absorption during the plastic phase. Therefore, architected materials with bending-dominated mechanical behavior have high crashworthiness and are appropriate for impact applications, such as helmets, shields, etc. [13, 14].

As it was aforementioned, the connectivity of an architected material's unit cell possesses a crucial role in the mechanical behavior of the structure. Therefore, for periodic struts and 2.5D architected materials, Maxwell established a mathematical condition, commonly known as Maxwell's stability criterion. Maxwell's stability criterion defines the mechanical behavior of an architected material (stretching-dominated or bending-dominated) based on the number of struts/beams (b) and nodes/joints (j) of each unit cell, as described by Eq. (4.13) and Eq. (4.14) [12, 15]. It is relevant to note that this classification, based on mechanical behavior, is only valid for architected materials with fixed nodes/joints. If the nodes are mobile, the structure behaves as a mechanism, and this criterion no longer applies. Figure 4.2c illustrates an indicative material diagram along with the curves of elastic modulus to density for both examined mechanical responses.

2*D* structures :

$$M = b - 2j + 3 \begin{cases} \text{if } M < 0 \text{ bending} - \text{dominated behavior} \\ \text{if } M \geq 0 \text{ stretching} - \text{dominated behavior} \end{cases} \quad (4.13)$$

3*D* structures :

$$M = b - 3j + 6 \begin{cases} \text{if } M < 0 \text{ bending} - \text{dominated behavior} \\ \text{if } M \geq 0 \text{ stretching} - \text{dominated behavior} \end{cases} \quad (4.14)$$

Finally, one notable class of architected materials is the auxetic lattice structures, such as the re-entrant and origami-like structures. These structures exhibit a stress–strain diagram similar to that of bending-dominated structures but possess the unique mechanical property of a negative Poisson ratio. A negative Poisson ratio is characterized by an increase and decrease of size in the dimensions perpendicular to the direction of loading during compression and tension, respectively. Furthermore, these architected materials can withstand large deformations, with low stress under loading, by folding the structure's elements. Due to these mechanical characteristics, auxetic materials are studied for 4D printing applications, where applied stresses (mechanical, thermal, etc.) result in significant changes in the structure's dimensions, enabling the modification of the entire structure's shape [16–18].

To conclude, for the characterization of the mechanical behavior of architected materials, it is essential to evaluate the main effective mechanical properties, such as the elastic modulus, yield stress, and peak stress, as well as the structure's Poisson ratio. These properties are typically determined through the employment of compressive quasi-static uniaxial tests for each applied relative density. Moreover, in order to calculate the constants of the scaling law for each effective property, tests are conducted on at least three different relative densities (necessary points for power laws) of the same architected material, and a curve-fitting process is applied. For instance, with a view to determining the mechanical behavior of an architected material, a minimum of three samples with varying relative densities are examined under quasi-static uniaxial compressive loading, and the effective mechanical properties are derived from the engineering effective stress–strain diagrams. Subsequently, a curve-fitting procedure is carried out for each effective mechanical property, based on the values for each relative density. The last important mechanical property of architected materials is the ability to absorb mechanical energy. In order to measure the energy absorption of an architected material, the surface area beneath the stress (σ) to strain (e) diagrams is computed up to 50% strain, before densification occurs, as described in the following equation, where W_V is the amount of the absorbed energy per volume typically measured MJ/m^3 or kJ/m^3 units. Moreover, the most widespread indicator of measuring this property is the specific energy absorption (SEA), which calculates the absorbed energy per mass (mainly J/g) and is expressed through Eq. (4.16). This process is derived from the corresponding international standards [19, 20].

$$W_V = \int_0^{\varepsilon_0} \sigma(\varepsilon)d\varepsilon \tag{4.15}$$

$$\text{SEA} = \frac{W_V}{\rho_{\text{material}}} \tag{4.16}$$

4.2.2   Numerical Evaluation

Besides the experimental methodology from the evaluation of architected materials, extensive research has been conducted in order to establish a numerical evaluation process for the capture or even the prediction of their mechanical response. The majority of these studies concern periodic architected materials due to the advantage of the structure's symmetry. The symmetric structure facilitates the solution of the mathematical problem employing representative volume elements (RVEs) and periodic boundary conditions (PBCs). Hence, the first step is the creation of RVEs for the desired architected material and then the examination of the mechanical behavior and anisotropy of the structure, numerically.

One of the most crucial aspects of architected material characterization is anisotropy which can lead to the degradation or even failure of a structure. Thus, the measurement of anisotropy, or the degree of directional dependence of mechanical properties, is critical in the analysis and characterization of newly created architected materials. This is particularly significant in commercial applications, where usually the properties of the materials should be similar across different directions. According to existing literature, [21, 22] two methods are commonly employed to measure anisotropy in architected materials, with the first method only applying to materials with a cubic external volume. This method calculates the Zener anisotropy ratio (A_r), which is a dimensionless number that quantifies the anisotropy of the material with values ranging from zero to one, with one indicating complete isotropy. The Zener ratio is calculated using the constants of the stiffness matrix (K) and the following equation, where C_{ij} are the elastic constants of the stiffness matrix, v is the structure Poisson ratio, G and E are the shear and elastic modulus, respectively.

$$K = \begin{bmatrix} C_{11} & C_{12} & C_{12} & 0 & 0 & 0 \\ C_{12} & C_{11} & C_{12} & 0 & 0 & 0 \\ C_{12} & C_{12} & C_{11} & 0 & 0 & 0 \\ 0 & 0 & 0 & C_{44} & 0 & 0 \\ 0 & 0 & 0 & 0 & C_{44} & 0 \\ 0 & 0 & 0 & 0 & 0 & C_{44} \end{bmatrix} \tag{4.17}$$

$$A_r = \frac{2(1+v)G}{E} = \frac{2 \cdot C_{44}}{C_{11} - C_{12}}. \tag{4.18}$$

The second method is suitable for both cubic and non-cubic architected materials and construction materials with linear elastic behavior, with the assistance of the E/E_{max} ratio. The E is the localized elastic modulus and the E_{max} is the maximum local elastic modulus. This method involves creating a surface map of the local elastic modulus for every point of the unit cell in all three directions, providing data on the location of the highest anisotropy within the structure. Through the contours of the surface map, the anisotropy of the structure is visible and regions of maximum and minimum normalized elastic modulus can be located. Additionally, Tancogne-Dejean et al. [22] have proposed the E_{max}/E_{min} ratio as another way to measure the anisotropy of a structure, which is a dimensionless number with a value of one for isotropic structures and a higher value indicating higher anisotropy. More specifically, the normalized elastic modulus and its contours were calculated and visualized by 3D plotting the directional effective elastic modulus (E_{abc}), according to the corresponding literature [23] and the following equation, where, $l = \cos(a)$, $m = \cos(b)$, and $n = \cos(c)$ with a, b, and c noting the angles from the three principal directions. Moreover, the constants S_{11}, S_{12}, and S_{44} represent the independent effective compliances. Figure 4.3 illustrates an indicative example of the surface map technique for sheet-Gyroid architected material.

$$\frac{1}{E_{abc}} = S_{11} - 2\left[(S_{11} - S_{12}) - \frac{1}{2}S_{44}\right](l^2m^2 + m^2n^2 + l^2n^2) \tag{4.19}$$

Another stage of the numerical methodology for architected materials is the proper set-up of PBCs. The vast majority of studies, which numerically examined the architected materials, utilize PBCs. PBCs consist of the boundary and symmetry conditions for a RVE. The symmetry conditions could be circular, mirrored, or linear depending on the applied use case. The most common case is the use of linear symmetry with periodic repetition of the architected materials RVE in the three

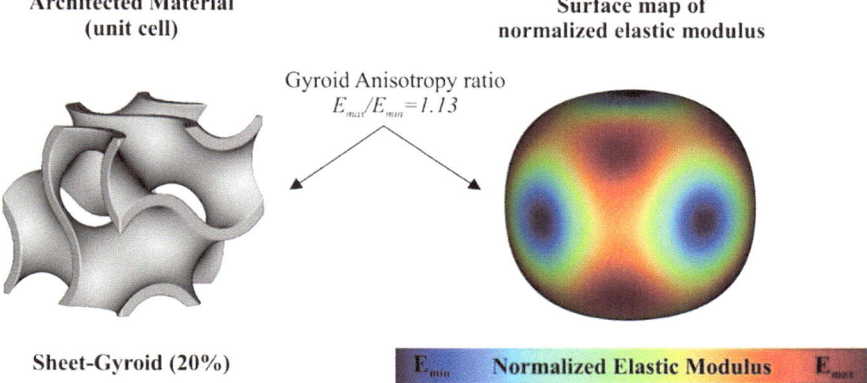

Fig. 4.3 Indicative example of surface map technique for sheet-Gyroid architected material

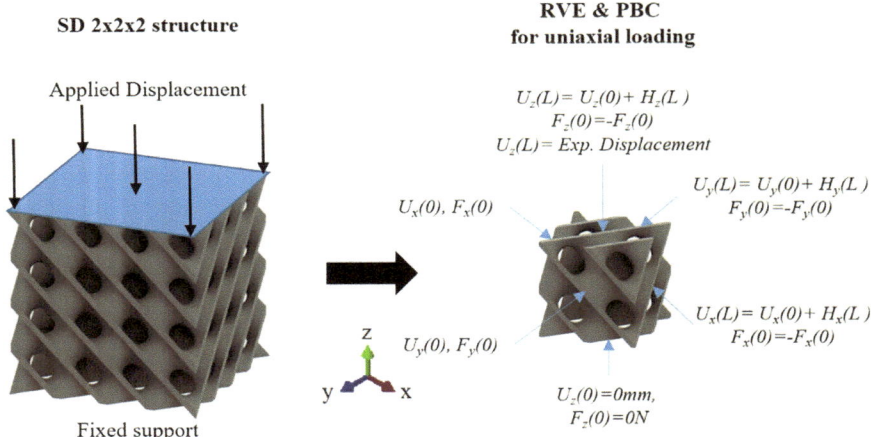

Fig. 4.4 PBCs of a 3D sheet Schwarz diamond RVE for uniaxial loading

dimensions with a period the employed length of the RVE. Figure 4.4 shows the proper PBCs of a 3D sheet Schwarz Diamond RVE for uniaxial compressive loading [24, 25] where H refers to the displacement gradients, L is the periodic length, U is the displacements, and F is the applied forces.

In order to sufficiently apply the aforementioned PBCs, proper preparation of RVEs should be performed. In detail, as it is depicted in Fig. 4.5, the design of architected material should be initially transformed to a tetrahedral element mesh with the desired element size and sharp edges, then it is modified to a volume mesh (solid domain with conform boundaries) and extracted as FE mesh in .cdb file format. This format is supported by various simulation software offering the necessary compatibility. Afterward, the file should be imported into the simulation software as an external model. The final step involves determining the appropriate angle threshold in order to produce a final CAD file with the necessary faces and edges. This is a high-importance process due to the fact that generates a light version of the CAD file, in contrast with the ordinary procedure which inserts the structure as a mesh with many elements and facets. In addition to this new procedure, an improved mesh is created in terms of structure and continuity.

Furthermore, due to the extensive degradation of the mechanical properties, elastic, plastic, viscoelastic, and hyper-elastic material models have been utilized across the existing published literature. The employed material model depends on the employed construction material and the applied relative density, as these two parameters influence both the degradation of the effective mechanical properties and the elastic/plastic behavior of the architected material.

Ordinary process

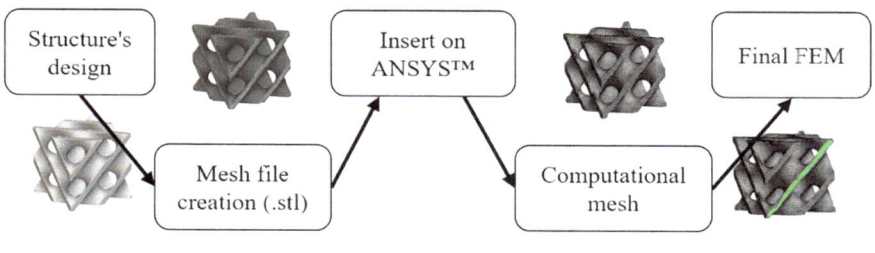

New integration process

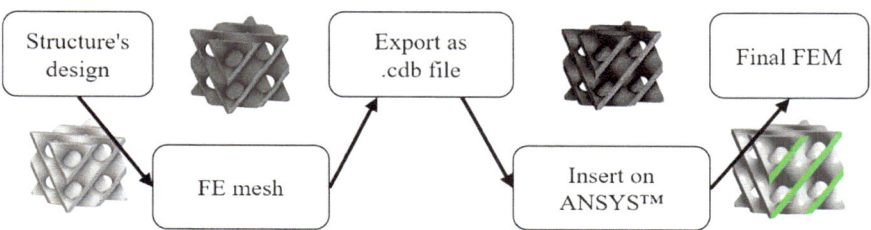

Fig. 4.5 Procedure of transferring architected materials geometry in ANSYS: ordinary process (top) and new integration process (bottom)

References

1. Tyflopoulos, E.; Flem, D. T.; Steinert, M.; Olsen, A. State of the Art of Generative Design and Topology Optimization and Potential Research Needs (2018).
2. O.M. Querin, G.P. Steven, Y.M. Xie, Evolutionary structural optimisation using an additive algorithm. Finite Elem. Anal. Des. **34**(3–4), 291–308 (2000). https://doi.org/10.1016/S0168-874X(99)00044-X
3. Y.M. Xie, G.P. Steven, *Evolutionary Structural Optimization.* (Springer London, London, 1997). https://doi.org/10.1007/978-1-4471-0985-3
4. X. Huang, Y.M. Xie, *Evolutionary Topology Optimization of Continuum Structures.* (Wiley, 2010). https://doi.org/10.1002/9780470689486
5. D. Tcherniak, Topology optimization of resonating structures using SIMP method. Int. J. Numer. Methods Eng. **54**(11), 1605–1622 (2002). https://doi.org/10.1002/nme.484
6. H. Jiao, Q. Zhou, S. Fan, Y. Li, *A New Hybrid Topology Optimization Method Coupling ESO and SIMP Method* (2015), pp. 373–384. https://doi.org/10.1007/978-3-662-44674-4_35
7. L.J. Gibson, M.F. Ashby, *Cellular Solids* (Cambridge University Press, 1997). https://doi.org/10.1017/CBO9781139878326
8. O. Al-Ketan, R. Rowshan, R.K. Abu Al-Rub, Topology-mechanical property relationship of 3D printed strut, skeletal, and sheet based periodic metallic cellular materials. Addit. Manuf. **19**, 167–183 (2018). https://doi.org/10.1016/j.addma.2017.12.006
9. O. Al-Ketan, R. Rezgui, R. Rowshan, H. Du, N.X. Fang, R.K. Abu Al-Rub, Microarchitected stretching-dominated mechanical metamaterials with minimal surface topologies. Adv. Eng. Mater. **20**(9), 1800029 (2018). https://doi.org/10.1002/adem.201800029

10. N. Kladovasilakis, K. Tsongas, I. Kostavelis, D. Tzovaras, D. Tzetzis, Effective mechanical properties of additive manufactured strut-lattice structures: experimental and finite element study. Adv. Eng. Mater. **24**(3), 2100879 (2022). https://doi.org/10.1002/adem.202100879
11. N. Kladovasilakis, K. Tsongas, I. Kostavelis, D. Tzovaras, D. Tzetzis, Effective mechanical properties of additive manufactured triply periodic minimal surfaces: experimental and finite element study. Int. J. Adv. Manufact. Technol. **121**(11–12), 7169–7189 (2022). https://doi.org/10.1007/s00170-022-09651-w
12. M.F. Ashby, The properties of foams and lattices. Philos. Trans. Royal Soc. A: Math., Phys. Eng. Sci. **2006**(364), 15–30 (1838). https://doi.org/10.1098/rsta.2005.1678
13. S.F. Khosroshahi, S.A. Tsampas, U. Galvanetto, Feasibility study on the use of a hierarchical lattice architecture for helmet liners. Mater. Today Commun. **14**, 312–323 (2018). https://doi.org/10.1016/j.mtcomm.2018.02.002
14. C.G. Ferro, S. Varetti, G. De Pasquale, P. Maggiore, Lattice structured impact absorber with embedded anti-icing system for aircraft wings fabricated with additive SLM process. Mater. Today Commun. **15**, 185–189 (2018). https://doi.org/10.1016/j.mtcomm.2018.03.007
15. M. Nasim, U. Galvanetto, Mechanical characterisation of additively manufactured PA12 lattice structures under quasi-static compression. Mater. Today Commun. **29**, 102902 (2021). https://doi.org/10.1016/j.mtcomm.2021.102902
16. C.M. González-Henríquez, M.A. Sarabia-Vallejos, J. Rodriguez-Hernandez, Polymers for additive manufacturing and 4D-printing: materials, methodologies, and biomedical applications. Prog. Polym. Sci. **94**, 57–116 (2019). https://doi.org/10.1016/j.progpolymsci.2019.03.001
17. X. Zheng, X. Guo, I. Watanabe, A mathematically defined 3D auxetic metamaterial with tunable mechanical and conduction properties. Mater. Des. **198**, 109313 (2021). https://doi.org/10.1016/j.matdes.2020.109313
18. L. Sun, W.M. Huang, Z. Ding, Y. Zhao, C.C. Wang, H. Purnawali, C. Tang, Stimulus-responsive shape memory materials: a review. Mater. Des. **33**, 577–640 (2012). https://doi.org/10.1016/j.matdes.2011.04.065
19. S.F. Fischer, Energy absorption efficiency of open-cell pure aluminum foams. Mater. Lett. **184**, 208–210 (2016). https://doi.org/10.1016/j.matlet.2016.08.061
20. M.M. Sychov, L.A. Lebedev, S.V. Dyachenko, L.A. Nefedova, Mechanical properties of energy-absorbing structures with triply periodic minimal surface topology. Acta Astronaut. **150**, 81–84 (2018). https://doi.org/10.1016/j.actaastro.2017.12.034
21. S. Al Hassanieh, A. Alhantoobi, K.A. Khan, M.A. Khan, Mechanical proper-ties and energy absorption characteristics of additively manufactured light-weight novel re-entrant plate-based lattice structures. Polymers (Basel) **13**(22), 3882 (2021). https://doi.org/10.3390/polym13223882
22. T. Tancogne-Dejean, X. Li, M. Diamantopoulou, C.C. Roth, D. Mohr, High strain rate response of additively-manufactured plate-lattices: experiments and modeling. J. Dyn. Behav. Mater. **5**(3), 361–375 (2019). https://doi.org/10.1007/s40870-019-00219-6
23. S. Khaleghi, F.N. Dehnavi, M. Baghani, M. Safdari, K. Wang, M. Baniassadi, On the directional elastic modulus of the TPMS structures and a novel hybridization method to control anisotropy. Mater. Des. **210**, 110074 (2021). https://doi.org/10.1016/j.matdes.2021.110074
24. F. Derveni, A.J. Gross, K.D. Peterman, S. Gerasimidis, Postbuckling behavior and imperfection sensitivity of elastic-plastic periodic plate-lattice materials. Extreme Mech. Lett. **50**, 101510 (2022). https://doi.org/10.1016/j.eml.2021.101510
25. H. Jia, H. Lei, P. Wang, J. Meng, C. Li, H. Zhou, X. Zhang, D. Fang, An experimental and numerical investigation of compressive response of designed Schwarz primitive triply periodic minimal surface with non-uniform shell thickness. Extreme Mech. Lett. **37**, 100671 (2020). https://doi.org/10.1016/j.eml.2020.100671

Chapter 5
Evaluation for Topology Optimization

5.1 Evaluation of Element-Based Approach

The evaluation of the element-based approach for topology optimization utilizing the SIMP algorithm focuses on its effectiveness in producing optimal structural designs with minimal material usage. This approach is advantageous due to its ability to refine material distribution at the element level (voxels), allowing for a precise and efficient optimization process. By applying the SIMP algorithm, the method ensures that the stiffness of the structure is maximized while adhering to specified constraints, such as volume or weight limits. The evaluation reveals that this approach effectively balances computational efficiency with the quality of the resulting topologies, making it a robust method for complex structural optimization problems. The iterative nature of the element-based approach tests via FEAs a vast number of possible material distribution scenarios and examines their mechanical response under certain operation conditions. The evaluation criteria for this topology optimization approach typically examine multiple aspects, such as structural performance, convergence, material distribution, computational efficiency, design complexity, manufacturability, robustness and sensitivity. The structural performance of the produced topologically optimized part is evaluated by maximizing the stiffness-to-weight ratio or minimizing compliance while adhering to specified constraints, such as volume or weight limits. The convergence criterion concerns the efficiency and reliability with which the algorithm converges to an optimal solution, including the rate of convergence and the stability of the process across iterations. The effectiveness of the material distribution, especially in algorithms like SIMP, should ideally penalize intermediate density values and encourage a clear distinction between solid and void regions. Regarding computation efficiency, due to the iterative nature of the process, it demands significant computational resources, especially for complex designs, for this reason, the computational efficiency of the element-based approach possesses a crucial role for its selection and applicability. Moreover, the level of detail and intricacy (design complexity) that the algorithm can achieve in the final topology,

N. Kladovasilakis et al., *Innovations in Topology Optimization*,
SpringerBriefs in Applied Sciences and Technology,
https://doi.org/10.1007/978-3-031-77700-4_5

including the resolution of fine features and the avoidance of numerical artifacts, is also one of the main criteria for an element-based approach. The practicality of the optimized design in terms of real-world fabrication, including considerations like minimum feature size, connectivity, and smoothness of boundaries for the produced topologically optimized design is one of the most crucial aspects that should be examined in this approach to ensure the production of a feasible design. Finally, the robustness and sensitivity of the element-based approach should be inspected in order to ensure process-independent results, in detail the approach's robustness is measured against variations in initial conditions, mesh dependency, and sensitivity to parameter choices, such as the penalization factor.

5.2   Evaluation for Architected Materials

In this section, the evaluation of the existing architected materials according to the literature and conducted research is presented. In detail, an in-depth analysis of the results from already published studies is demonstrated for several architected materials fabricated with various AM technologies. Furthermore, the most crucial characteristics and properties of the architected materials along with their evaluation methodologies, which were presented in previous sections, have been referred to. It is worth mentioning that the presented results concern both numerical and experimental research studies.

Mechanical behavior:

The first and most important research aspect of all conducted studies for the subject of architected materials' evaluation, is classifying the structure's mechanical behavior as either stretching-dominated or bending-dominated, and the extraction of the corresponding scaling laws. Both of these elements are crucial for further investigation, as it was discussed in previous subchapters. Scaling laws are essential tools for the evaluation of architected materials, as they describe the overall mechanical behavior and express the effective mechanical properties based on the employed relative densities. Moreover, they can be directly linked to design-related parameters. As it was mentioned, the scaling laws are applicable to architected materials with relative densities below 50% and the most essential scaling law equations are the equations regarding the effective elastic modulus and the effective yield strength (Eq. 4.9–Eq. 4.12), due to the fact that the values of their exponents are indicative for the mechanical behavior of the structure. According to the literature [1, 2], the values of the constant G_E could range from 0.01 to 4, and the constant n is a positive number. In all published studies, the extraction of the scaling laws constants is a process performed through compression uniaxial quasi-static tests on lattice specimens. Therefore, knowing the exponential trend of scaling laws, the minimum number of examined relative densities is three to export produce a proper curve and calculate the constants G and n.

A plethora of experimental published studies have examined the mechanical response of various architected materials, mainly 3D strut and 3D sheet TPMS structures, made of numerous construction materials, and extracted the scaling laws using the aforementioned methodology. Toward this direction, Fig. 5.1 graphically illustrates results from published studies of the experimental effective elastic modulus and effective yield strength for numerous architected materials and applied relative densities [3–7]. Also, Fig. 5.1 depicts the following: the structures' identity, their construction material, and the nominal slope of the stretching-dominated and bending-dominated behaviors. It is clearly observed that there is a declining pattern in the elastic modulus and strength of architected materials as the relative density decreases. This trend is intense for structures with bending-dominated behavior, especially in Fig. 5.1a, where the two distinct behaviors have a larger difference in slopes. Furthermore, Fig. 5.1b shows an enhanced mechanical strength for the stretching-dominated structures for the same relative densities. In addition, the 3D sheet TPMS architected material, namely the sheet Gyroid and Diamond, are located on the top region of the effective strength diagram (Fig. 5.1b) indicating the superior mechanical strength of these architected materials. Finally, the 3D strut architected material could show intense stretching-dominated behavior, such as Octet structures, however, their mechanical strength is significantly lower. It is worth noting that the plotting of these graphical charts is possible only for the representation of architected materials fabricated with similar construction materials (i.e. only polymer or only metals) in order to obtain clear and comparable results.

Figure 5.1 presents the higher structural integrity of 3D TPMS architected materials in terms of structural integrity. Moreover, the 3D strut and TPMS architected materials are in the upper region of the foam's bubble in the chart, showing an advanced mechanical performance, as it is portrayed in Fig. 5.1c. In addition, it is worth mentioning that the architected material can exhibit similar mechanical behavior to elastic and elastomeric material with significantly lower density (weight), providing the potential for replacing ordinary bulk polymers and elastomers in energy absorption applications.

Thus, in the last five years, many studies have focused on the investigation and examination of these structures, including the research of this book. To obtain a more sufficient image of the TPMS structure, Table 5.1 summarizes the results of existing published research [8–15]. These studies have investigated the TPMS architected materials with the highest potential using several construction materials, both polymers and metals. Hence, Table 5.1 lists the constants and the exponents of the scaling laws for the effective elastic modulus and the effective yield strength coupled with the applied architected materials and the employed construction materials. Knowing the print-out material elastic modulus and yield strength is easy to calculate the aforementioned properties for every relative density through the equation from the Sect. 3.3.

According to Table 5.1, most of the 3D sheet TPMS showed a mechanical behavior closer to the stretching-dominated along with enhanced yield strength. This occurred due to the sufficient connectivity of the 3D sheets in the structures. On the other hand, 3D strut TPMS architected materials exhibited closer to the

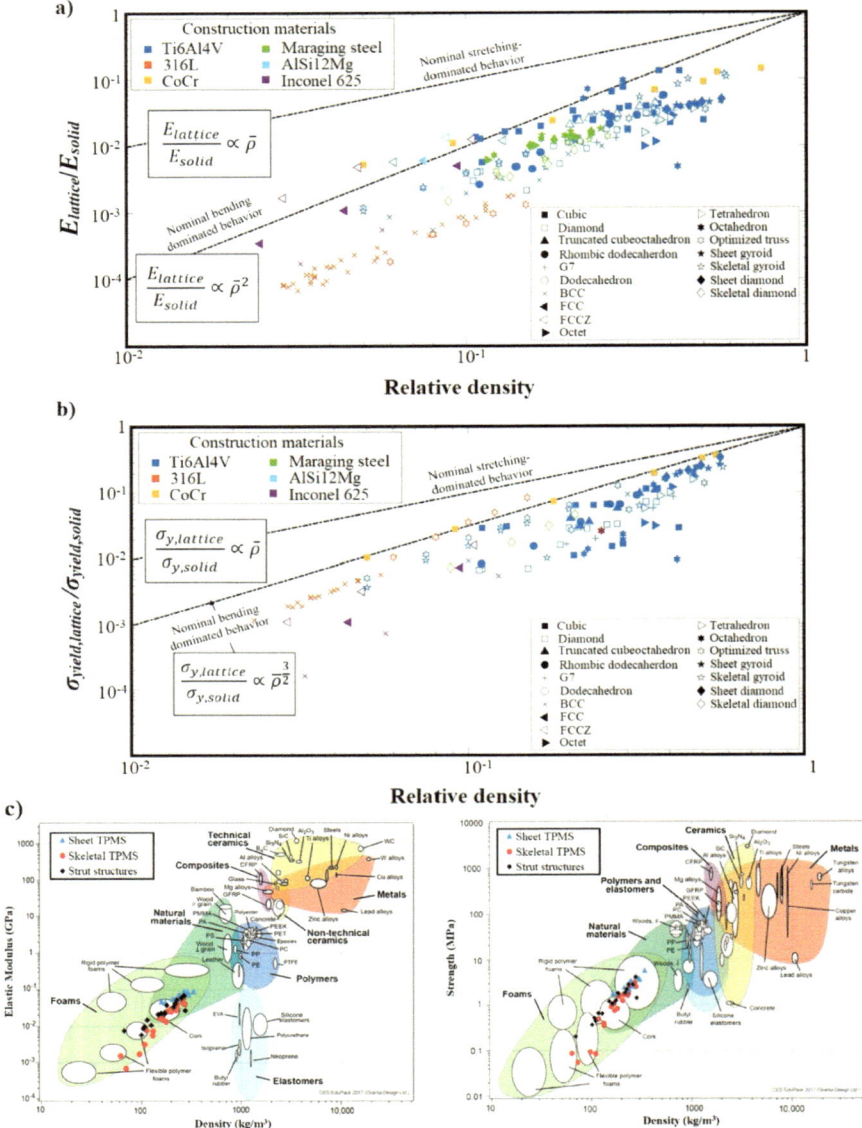

Fig. 5.1 Graphical relation of relative density and: **a** Effective elastic modulus, **b** Effective yield strength for several architected materials constructed with metal SLM AM technologies, **c** Representative bubble diagrams for elastic modulus (left) and strength (right) of the position of architected material regarding the other bulk material categories [3]

Table 5.1 Scaling laws constants for various architected materials constructed with several materials [18]

Construction material	TPMS architected material	Elastic modulus (GPa)		Yield strength (MPa)	
		G_E	n_E	G_σ	n_σ
Maraging steel	Sheet-primitive	0.11	1.31	1.43	1.77
	Sheet-diamond	0.04	0.522	0.93	1.39
	Sheet-gyroid	0.11	1.23	0.89	1.43
	Sheet-IWP	0.09	1.15	2.35	2.13
Stainless steel	Sheet-primitive	0.56	1.89	2.67	2.23
	Sheet-diamond	0.43	1.42	1.84	1.76
	Sheet-gyroid	1.14	2.23	2.74	2.1
Titanium	Sheet-primitive	0.11	1.12	0.81	1.80
	Sheet-diamond	0.09	0.79	0.39	1.30
	Sheet-gyroid	0.12	1.1	0.85	2.10
	Sheet-IWP	0.10	1.03	0.73	1.90
PA2200	Sheet-primitive	0.34	1.52	1.51	1.75
	Sheet-IWP	0.42	1.18	1.83	1.65
	Sheet-neovius	0.83	1.59	2.8	1.89
Maraging steel	Skeletal-IWP	0.19	2.01	1.53	2.17
	Skeletal-diamond	0.35	2.22	4.42	2.73
	Skeletal-gyroid	0.14	1.68	1.19	1.86
Aluminum	Skeletal-gyroid	0.42	2	8.98	2.55
PA11	Skeletal-diamond	0.55	2.24	1.15	2.48
	Skeletal-IWP	0.83	2.71	0.82	2.56
	Skeletal-gyroid	0.82	2.36	0.57	2.10
	Skeletal-Fischer & Koch	0.91	2.63	0.98	2.6
PA12	Sheet-Schwarz primitive	0.4	2.01	1.61	2.55
	Sheet-gyroid	0.16	1.22	0.34	1.14
	Sheet-Schwarz diamond	0.43	1.2	0.37	0.91
	Sheet-Neovious	0.42	1.2	0.34	0.77

bending-dominated mechanical response. The main reason for this performance is the bending of the curved beams of the material's structures due to their response to follow the TPMS's unique curvature. More specifically, Al-Ketan et al. [9] showed that the 3D sheet TPMS Diamond architected material exhibited intense stretching-dominated in contrast with the corresponding skeletal structure which revealed clear bending-dominated behavior. In addition, the degradation of the effective mechanical properties occurs milder in stretching-dominated behavior, indicatively for the same structure (Diamond), the yield strength is at 10% of construction material's yield strength for the sheet structure in contrast with the 5.46% for the skeletal structure at a relative density of 20%. The same trend is observed for Gyroid structures (20% relative density) with 9% for sheet configuration and 6% for skeletal one. Furthermore, it is evident that the values of the constants depend on both the applied construction and the applied architected materials. For strong metal construction materials, the architected materials tend to follow a stretching-dominated behavior despite the applied geometry. However, as the construction material becomes softer, such as Aluminum alloys [14], or even more such as a rigid polymer (PA11, PA12) [1, 15, 16] the examined architected materials tend to reveal a more bending-dominated behavior due to the increase of the effective elastic modulus exponent. An extreme condition of this situation is the employment of elastomeric construction materials in the architected material's structure revealing strictly bending-dominated behavior during their characterization. An indicative example of this is the evaluated constants of the skeletal TPMS for Abou-Ali et al. [15], which showed greater bending-dominated behavior compared with Al-Ketan's [17] study for the same construction due to the employment of softer construction material.

Numerical Evaluation:

Hence, the first step was the creation of RVEs for the desired architected material and then the examination of the anisotropy of the structure. As mentioned in previous chapters, two methods are employed for the anisotropy measurements. The most common is with the Zener ratio, however, the Zener ratio is applicable only to cubic RVEs. Thus, recent studies have suggested the utilization of normalized elastic modulus. According to existing studies [19,20, 38] Zener ratio (A_r) is a mechanical property that also follows a scaling law (Eq. 5.1), but it has different values of constants C_z and n_z for different architected materials, lacking a clear pattern of anisotropic behavior. In addition, from Eq. (5.1), it is obvious that architected materials with higher relative densities reveal a more elastic-isotropic behavior than low-density structures. Moreover, the second method of anisotropy has gained increasing interest in recent years due to the ease of calculation, the visualization of the anisotropy, and the lack of geometric restrictions. Therefore, the normalized elastic modulus and the E_{max}/E_{min} ratio consist of a useful tool for the evaluation of anisotropy [21, 22] which can be performed with polar diagrams or with the overall surface maps of the E_{max}/E_{min} ratio, as it is depicted in Fig. 5.2. For the definition of this method, it is derived that an elastic-isotropic behavior is achieved when E_{max}/E_{min} ratio is equal to one, in all the other cases $E_{max}/E_{min} > 1$. According to the literature [20, 38] architected materials with stretching-dominated behavior tend to reveal lower anisotropy than bending-dominated structures. Also, it is worth noting that for non-symmetric architected materials an extensive anisotropy could occur, as it is shown in Fig. 5.2, with the E_{max} being more than 10 times greater than the E_{min}.

$$A_r = C_z \cdot (\overline{\rho})^{n_z} \tag{5.1}$$

Therefore, in Table 5.2 the values of the E_{max}/E_{min} and the Zener ratios are presented for each examined architected material and in every relative density. As it is shown in the above-listed table, Gyroid and Kelvin structures experience a

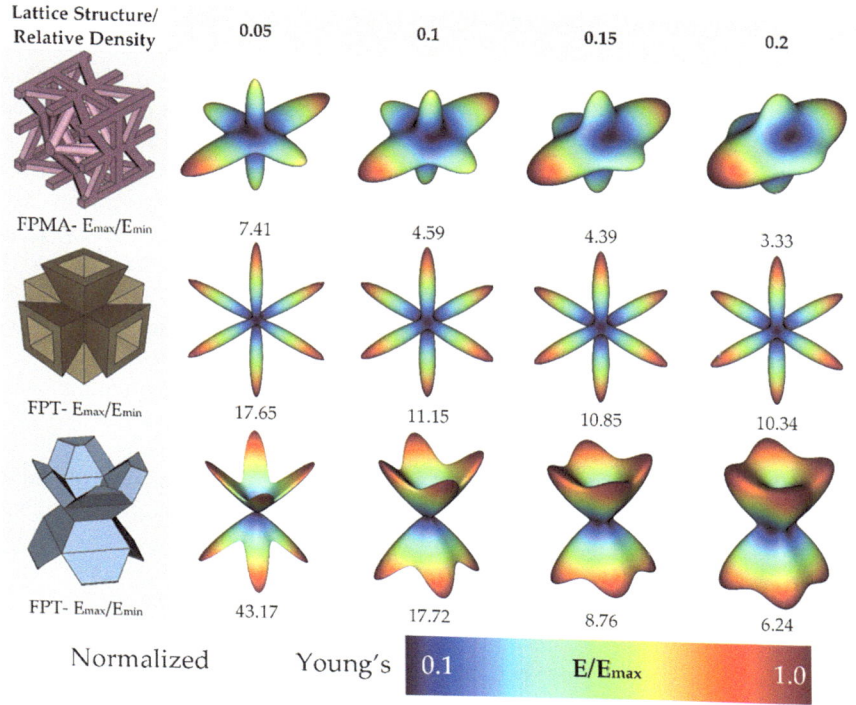

Lattice Structure/ Relative Density	0.05	0.1	0.15	0.2
FPMA- E_{max}/E_{min}	7.41	4.59	4.39	3.33
FPT- E_{max}/E_{min}	17.65	11.15	10.85	10.34
FPT- E_{max}/E_{min}	43.17	17.72	8.76	6.24

Normalized Young's | 0.1 E/E_{max} 1.0

Fig. 5.2 Indicative surface maps and E_{max}/E_{min} ratio for non-cubic structure in different relative densities [20]

rare behavior with the rise of the anisotropy ratios as the relative density increases, however, this rise is negligible and mainly occurs due to the non-uniform distribution of the material. On the other hand, all the other architected materials revealed a reduction in their anisotropy as their relative density increased. The best performance in terms of anisotropy was observed from the Gyroid structure for the relative densities. However, the SD structure exhibited a great improvement in anisotropy ratios as relative density increased resulting in similar anisotropy with Gyroid at 30% relative density. Moreover, the architected materials with the higher anisotropy were the RD structure and followed by the Neovius structure due to the high concentration of material in specific regions of the structure's volume. It is worth mentioning that, overall, the examined architected materials showed relatively low anisotropy without extreme values (below 3), for both ratios and decreased further, below the value of two, as the relative density increased. Furthermore, the surface maps of normalized elastic modulus for every examined architected material at the indicative relative density of 20%. In this graphical illustration, the strongest and weakest regions of the architected material's structures can be observed. The strongest regions are illustrated with red color and indicate the high levels of normalized elastic modulus, in which the structures have the highest stiffness. In contrast, the blue color shows the areas with the lowest normalized elastic modulus and the lowest stiffness resulting in poor mechanical performance.

Table 5.3 presents a series of material models employed in different published studies for both 3D strut and TPMS architected materials, aimed at capturing their mechanical response when constructed from different materials [12, 23–32]. It is pertinent to note that when an architected material is made from a polymer, the degradation of its effective mechanical properties reaches levels comparable to elastomeric materials, especially in terms of the effective elastic modulus. This results in a mechanical response similar to a stiff elastomer material. Thus, in this book and in other recent studies, elastic-viscoplastic and hyper-elastic FE material models have been utilized to accurately capture the non-linear response and the post-yielding plateau of 3D printed polymeric architected materials. These models can also be employed to extract effective mechanical properties and corresponding scaling laws. However, the acquisition of experimental data is eventually essential in order to validate and verify the numerical results obtained from these FE models.

Table 5.4 presents the constants of the scaling laws for effective elastic modulus and yield strength, as they were calculated via numerical analysis for different construction and architected materials. As it was expected, the numerical results are close enough to the experimental data. For comparison reasons, similar architected materials with Table 5.1, are listed in Table 5.4. For example, for 3D sheet Schwarz Primitive architected materials made of polymer, similar values of constants were evaluated either from the experimental or the numerical procedure. In addition, Fig. 5.3 illustrates an example of a numerical simulation of the mechanical response of the Schwarz Primitive architected material under compression loads. This figure

Table 5.2 Values for E_{max}/E_{min} and Zener ratios for eight architected materials in different relative densities

Architected materials	Relative density					
	10%		20%		30%	
	E_{max}/E_{min}	Zener ratio	E_{max}/E_{min}	Zener ratio	E_{max}/E_{min}	Zener ratio
Octet	1.703	1.862	1.615	1.748	1.524	1.635
Rhombic dodecahedron	2.661	2.773	2.206	2.362	1.795	1.927
Kelvin	1.308	1.322	1.391	1.447	1.366	1.453
Weaire-Phelan	2.135	2.198	1.889	1.938	1.716	1.742
Schwarz primitive	2.826	3.13	2.217	2.423	1.773	1.904
Gyroid	1.087	1.1	1.122	1.138	1.114	1.16
Schwarz diamond	1.436	1.497	1.257	1.294	1.113	1.127
Neovious	2.556	2.841	1.83	1.972	1.48	1.558

Table 5.3 Numerical research with different applied FE material models for indicative construction and architected materials [18]

Material models	Construction materials	Architected materials	
		Strut	Sheet
Elastic [23–25]	PA2200, polymer composites	✓	✓
Elastic–plastic [26, 27]	Titanium alloys Photopolymer resins	✓	✓
Elastic-viscoplastic [28, 29]	Ti6Al4V, PA2200	✓	✓
Viscoelastic [30]	Polymer	–	✓
Hyper-elastic [1, 12, 16]	PA2200, PA	✓	✓
Modular dynamics [31, 32]	Graphene foams and crystalline structures	✓	✓

demonstrates that the appropriate finite element material models can effectively simulate the actual mechanical behavior of the specimen, both numerically (via stress–strain diagrams) and visually (through images from compression experiments and FEAs) [1].

Energy Absorption and Crashworthiness

Architected materials are a class of materials engineered with a specific arrangement of microstructures that provide tailored mechanical and thermal properties. Hence,

Table 5.4 Numerically obtained scaling laws for different architected materials [18]

Construction material	TPMS architected material	Elastic modulus (GPa)		Yield strength (MPa)	
		G_E	n_E	G_σ	n_σ
Polymer	Sheet-primitive	0.56	1.52	–	–
	Sheet-Fischer & Koch	0.71	1.24	–	–
	Sheet-Neovius	0.70	1.22	–	–
	Sheet-IWP	0.56	1.41	–	–
Photopolymer resin	Sheet-diamond	0.55	1.25	0.58	1.17
	Sheet-gyroid	0.51	1.38	0.44	1.24
	Octet	0.30	1.28	0.32	1
	Skeletal-diamond	0.49	2.04	0.57	1.84
	Skeletal-gyroid	0.81	2.03	0.44	1.60
Maraging steel	Sheet-IWP	0.79	1.48	0.69	1.24
	Skeletal-IWP	1.28	2.64	1.05	2.14
Polymer	Sheet-primitive	0.24	1.17	0.49	1.14
	Skeletal-primitive	0.61	1.57	0.79	1.36
PA2200	Sheet-gyroid	0.62	1.27	–	–

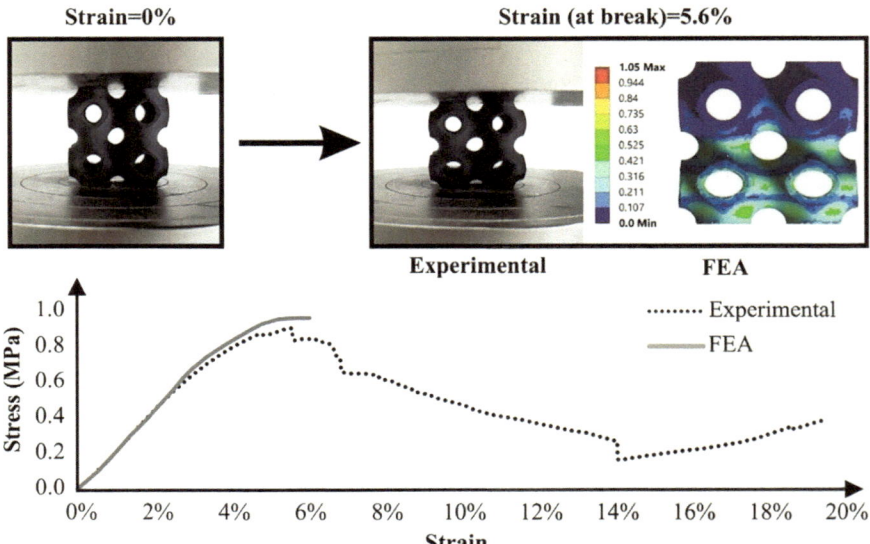

Fig. 5.3 Indicative images of comparison of experimental data and FEA's results for $2 \times 2 \times 2$ specimens for Schwarz Primitive [1]. (Copyrights Springer Nature)

architected materials can be designed to have high energy absorption and crashworthiness capabilities, making them attractive for applications in impact protection, energy dissipation, and shock mitigation. According to the literature [34], architected materials have proven superior energy absorption capabilities compared to traditional materials, and their design and engineering can be tailored to meet specific energy dissipation requirements. Static energy absorption is measured through quasi-static tests, and as such, it enables a structure to absorb energy in static phenomena. As it was mentioned in previous chapters, the energy absorption of a structure is calculated through the surface area below the stress–strain diagram for compression loading. SEA is utilized as an indicator value of absorbed energy. Figure 5.4a portrays the SEA in a bubble chart for different 3D strut and sheet-network architected materials constructed with several materials [33–38]. Finally, the efficiency of the absorbed energy (W_e) for an architected material could be measured according to Fig. 5.4b and Eq. (5.2). In detail, the efficiency is calculated via the ratio of the structure's absorbed energy (W_b) to the theoretical energy absorption of an ideal absorber.

$$W_e = \frac{W_b}{W_a + W_b} \cdot 100[\%] \rightarrow W_e = \frac{W_b}{\sigma_{\text{peak},50\%} \cdot \varepsilon_{50\%}} \cdot 100[\%] \qquad (5.2)$$

Moreover, Table 5.5 lists SEA for 3D TPMS architected material constructed from various materials [7, 9, 11–15]. As it was expected, architected materials with a higher relative density exhibit higher SEA. According to existing literature, as shown in Fig. 5.4a and Table 5.5, SEA values can vary widely-from 0.08 J/g for ultra-low

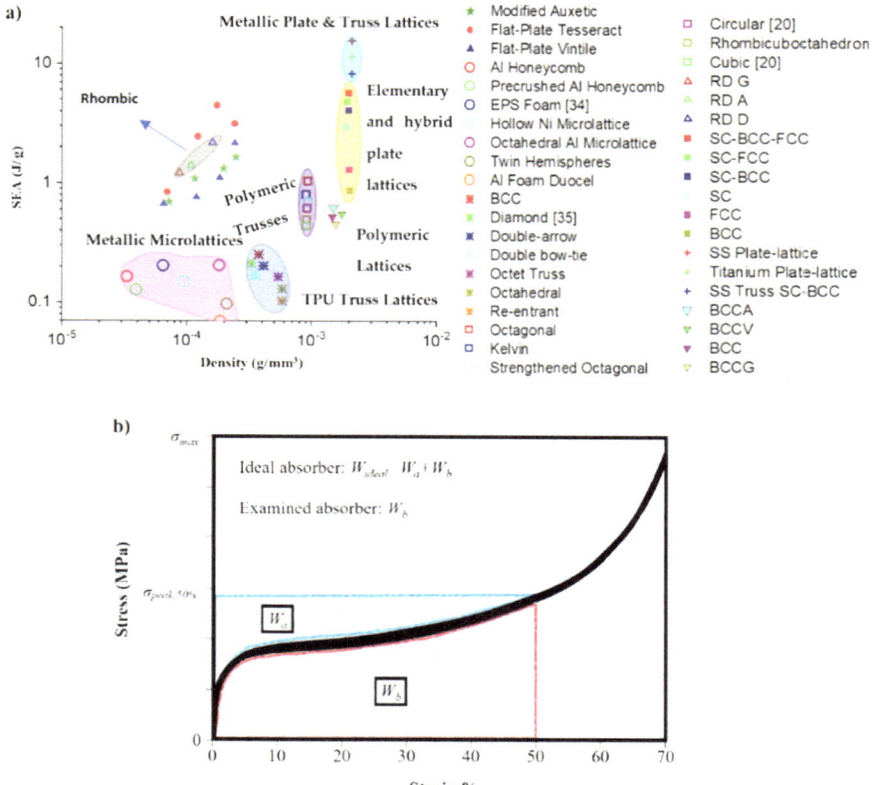

Fig. 5.4 **a** SEA in a bubble chart for different 3D strut and sheet-network architected materials constructed with several materials [36], **b** Stress–strain diagram and energy absorption regions for a theoretical ideal absorber and an indicative examined structure

density metallic micro-lattices and for TPU-made structures, to approximately 31 J/g for metal 3D sheet TPMS architected materials, such as SD structure (with ≈ 30–35% relative density). In contrast, 3D strut architected materials demonstrated the lowest energy absorption performance, with SEA ranging from 0.08 J/g to nearly 1 J/g. On the other hand, the 3D interconnected plate architected materials had improved energy absorption capabilities with SEA between 0.7 J/g to 11 J/g depending on the construction material. Regarding the TPMS architected materials, their overall performance in terms of energy absorption is superior to the other structures. More specifically, the skeletal TPMS structures revealed values of SEA between 0.45 J/g to 10.06 J/g. In addition, the most advanced energy absorption performance was observed for the sheet-TPMS architected materials with values of SEA ranging from 2.47 J/g to 30.89 J/g. From all these results, it is obvious that besides the architected materials' topology, the construction material possesses a fundamental role in the energy absorption capabilities of a structure. Figure 5.5 illustrates the energy efficiency of eight indicative and widely used architected materials (four strut-based and four sheet-TPMS) [1, 16], revealing that the TPMS-based architected materials achieved high energy efficiency values, ranging from 41% to nearly 68%, making them suitable for impact and energy applications.

Crashworthiness is a measure of a structure's ability to handle the dynamic phenomena of a crash or impact event [39]. More specifically, it refers to the capacity of a structure, such as a vehicle, aircraft, or building, to absorb energy from dynamic loading (i.e. high-velocity impacts, etc.) and reduce the severity of the event [40]. The crashworthiness of a structure, especially of an architected material, has high value for industrial and commercial applications, and due to this reason, it has been investigated extensively during the last few years. The main difference between an impact and a quasi-static loading is the speed of the event, i.e. the strain rate. Hence, the nature of an experiment depends on the applied strain rate of the loading. As it is shown in Fig. 5.6a, different strain rates result in different analyses and utilize different test equipment [41]. For example, for strain rate up to $10s^{-1}$, the phenomenon is similar to static/quasi-static behavior and the conventional energy

Table 5.5 SEA for different 3D TPMS architected materials constructed with several materials [33–38]

Construction material	TPMS architected material	Density (g/mm^3)	SEA (J/g)
Maraging steel	Sheet-primitive	$0.8 \times 10^{-3} - 1.6 \times 10^{-3}$	$6.33 - 11.8$
	Sheet-diamond	$1.3 \times 10^{-3} - 2.2 \times 10^{-3}$	$13.5 - 17.53$
	Sheet-gyroid	$1 \times 10^{-3} - 1.9 \times 10^{-3}$	$8.8 - 14.53$
	Sheet-IWP	$1.3 \times 10^{-3} - 1.8 \times 10^{-3}$	$10.15 - 14.53$
PA11	Sheet-diamond	$0.2 \times 10^{-3} - 0.4 \times 10^{-3}$	$5.46 - 7.72$
	Sheet-IWP	$0.2 \times 10^{-3} - 0.3 \times 10^{-3}$	$2.47 - 4.16$
	Sheet-gyroid	$0.2 \times 10^{-3} - 0.4 \times 10^{-3}$	$5.16 - 6.86$
SS 316L	Sheet-gyroid	$1.1 \times 10^{-3} - 2.9 \times 10^{-3}$	$10.93 - 21.3$
	Sheet-primitive	$0.8 \times 10^{-3} - 2.7 \times 10^{-3}$	$6.2 - 18.19$
	Sheet-diamond	$1.4 \times 10^{-3} - 2.5 \times 10^{-3}$	$22.01 - 30.89$
Maraging steel	Skeletal-IWP	$0.7 \times 10^{-3} - 1.7 \times 10^{-3}$	$2.82 - 7.4$
	Skeletal-diamond	$1 \times 10^{-3} - 2.1 \times 10^{-3}$	$1.2 - 6.66$
	Skeletal-gyroid	$1. \times 10^{-3} - 1.8 \times 10^{-3}$	$5.53 - 10.06$
PA11	Skeletal-IWP	$0.1 \times 10^{-3} - 0.3 \times 10^{-3}$	$0.53 - 4.35$
	Skeletal-gyroid		$0.9 - 4.8$
	Skeletal-diamond		$0.45 - 7.37$
PA12	Sheet-Schwarz primitive	$0.1 \times 10^{-3} - 0.3 \times 10^{-3}$	0.198
	Sheet-gyroid		0.533
	Sheet-Schwarz diamond		0.951
	Sheet-neovious		0.903

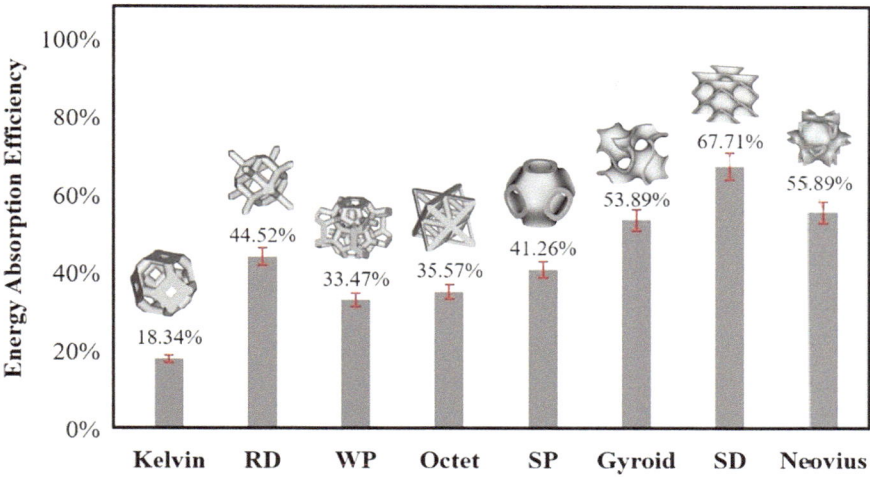

Fig. 5.5 Energy absorption efficiency of each examined architected material [1, 16]

absorption can be measured. On the contrary, if the strain rate is above $10^2 s^{-1}$, the test is characterized as a dynamic phenomenon (impact, crash, exposition, etc.), and the crashworthiness of the structures is measured with the same index, i.e. the SEA. The differentiation of energy absorption behavior for a structure based on the applied strain rate mainly occurs due to the strain rate hardening. Strain rate hardening, also known as strain rate sensitivity, is a phenomenon that occurs in structures subjected to high strain rate deformation, where the strength and hardness of the construction material increase as the strain rate increases. The mechanism of strain rate hardening is attributed to dynamic recovery, dynamic recrystallization, and grain refinement in the material's microstructure. The increased temperature and stress generated by high strain rate deformation result in the movement of dislocations and the formation of new grain boundaries, leading to a change in the microstructure and increased strength and hardness. Therefore, for high strain rates, the absorbed energy is significantly higher than the absorbed energy for low strain rates ($W_{HS} \gg W_{LS}$). These phenomena are also applicable to the energy absorption and crashworthiness of the architected materials [42]. More specifically, Novak et al. [42] have shown that SEA and peak strength could be increased up to 12.3% and 16.4%, respectively, for high strain rate loading at the same specimens (applied geometry and relative density). This incensement of the energy absorption and strength is the quantification of the strain rate hardening, due to the construction material's hardening and the applied deformation mode due to the inertia effect. It is worth explaining the observed deformation mode under the high strain loading and comparing it with the deformation mode in quasi-static loading. In quasi-static loading, the received loads on the specimens are distributed in the whole architected material's structure leading to a more uniform stress distribution and the development of intense shear stresses. Alternatively, in high-strain loading, the applied loads concentrate at the first unit cell layer, which is in contact with the moving plate, due to inertia. This effect results in the layer-by-layer destruction of the specimen. Moreover, when a layer is compressed rapidly with significant loads, the hardening of the construction material is observed with the aforementioned microstructure mechanisms. This phenomenon happens for each unit cell's layer of the structure resulting in a sufficient increase of the strength and SEA. Figure 5.6b depicts an indicative comparison stress–strain diagram for quasi-static and high-strain loading coupled with the measured SEA and the images from the observed plastic strain for each case.

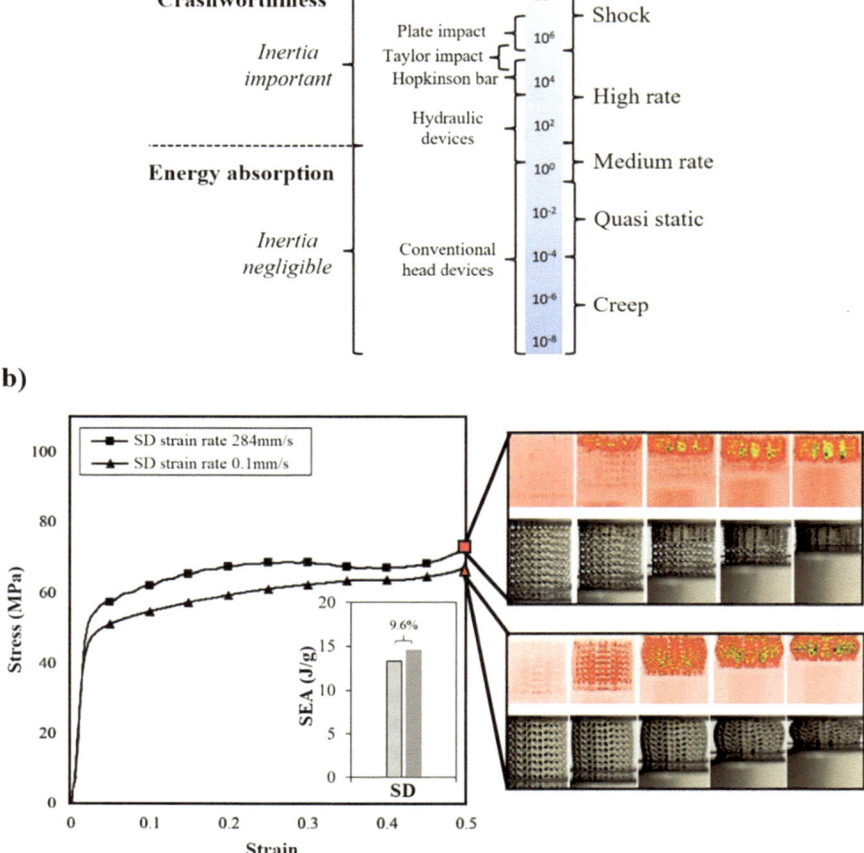

Fig. 5.6 **a** Approximate classification of energy absorption tests based on strain rates (in s^{-1}), **b** Comparison diagrams for SD structure made of SS316L at 21% relative density along with their deformation modes [42] (Copyrights 2018 Elsevier)

5.3 Evaluation of Optimized Architected Materials

The evaluation of optimized architected materials is similar to the evaluation process of original architected materials. Both structures obey scaling laws regarding the applied overall relative density. In this context, the authors of this book have investigated three distinct optimization methods for architected materials. The first method was the creation of entirely new hybrid architected materials, the second method was the functional gradation of relative density and the third method was the use of IPCs. In detail, in the context of this book, four novel and uniquely architected materials were developed and examined by hybridizing pairs of the above-studied architected materials and creating new cellular structures. Then, the optimization method of functional gradation of the relative density was examined for both the existing and the new hybrid material. Finally, the method of interpenetrating phase composites utilizing architected materials was investigated through both literature and experimental aspects, focusing mainly on energy absorption applications.

5.3.1 Hybrid Architected Materials

The idea involves the examination of the mechanical behavior of two different cellular materials to create a new hybrid unit cell with superior mechanical performance. Specifically, the research aimed to develop four innovative hybrid architected materials [43]. The main concept is to reduce the stress concentration regions and achieve more uniform stress distributions in the new design's volume. In Fig. 5.7, the hybridization process is depicted. It is worth noting that this methodology can be also implemented in other architected materials to produce entirely new hybrid architected materials.

Through the current study, four different hybrid cellular materials were designed and examined. These materials combined the Schwarz Primitive and Kelvin structures (SP&K), the Kelvin and Rhombic Dodecahedron structures (K&RD), the Neovius and Rhombic Dodecahedron structures (N&RD), and the Schwarz Diamond and Face-Centered Cubic structures (SD&FCC). The influence of strut/wall thickness on relative densities was investigated for these developed structures, revealing an exponential relationship for the K&RD structure and an almost linear relationship for the others. The anisotropy of these lattices was analyzed using normalized elastic modulus diagrams and E_{max}/E_{min} ratios, showing that SP&K had the highest anisotropy, while the other lattices exhibited nearly negligible anisotropy. The mechanical behavior of the hybrid lattices was examined under quasi-static compression at relative densities of 10%, 20%, 30%, and 40%. The SP&K and K&RD displayed bending-dominated behavior, whereas the N&RD and SD&FCC structures exhibited mild and intense stretching-dominated behavior, respectively. Scaling laws for each hybrid were plotted and quantified using the acquired data.

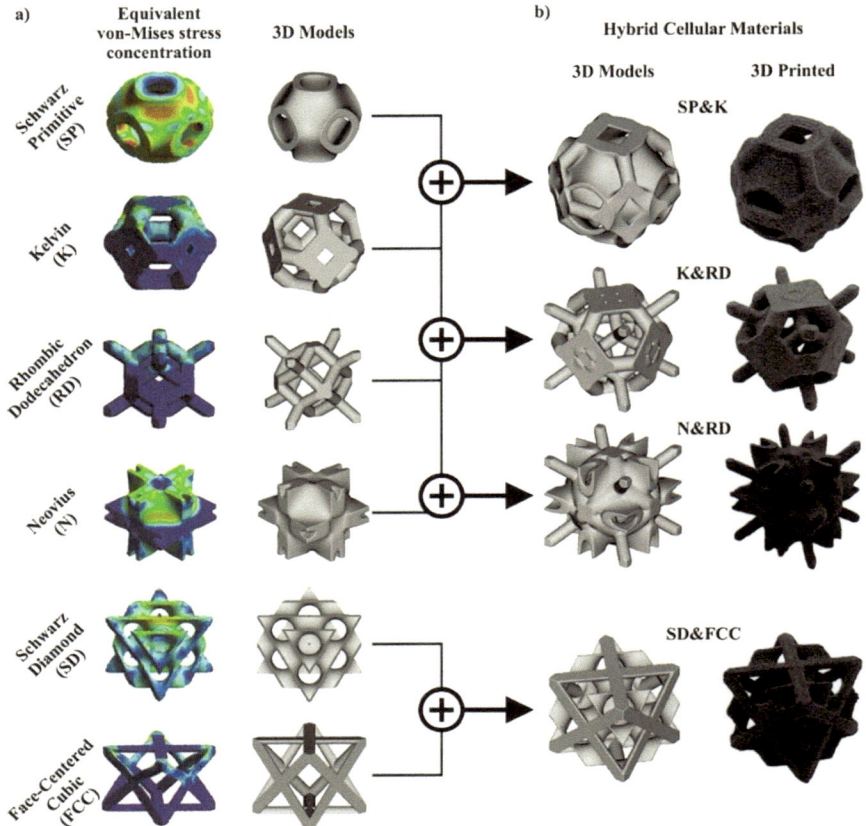

Fig. 5.7 Equivalent von-Mises stress and 3D models of the selected structures [43] (Copyrights 2023 Elsevier)

Finite Element (FE) models were developed for each structure to match the experimental data. These models accurately simulated the mechanical response of the lattices, revealing stress concentration regions and stress distribution. Verification specimens were designed, manufactured, and tested under compression to validate the reliability of the developed FE models and to extract the mechanical behavior of these lattices for more complex structures. The Finite Element Analysis (FEA) results closely matched the experimental data, demonstrating the efficiency and accuracy of the developed models. The combination of experimental and FEA data showed stress distribution during low strain rate tests, with fractures occurring at the top side of the specimens. Finally, the experimental results indicated that the hybrid architected materials exhibited enhanced mechanical strength compared to the original lattices. This improvement was attributed to the more uniform stress distribution within the volume and the avoidance of intense stress concentration in specific regions. Figure 5.8 summarizes the findings of the conducted research.

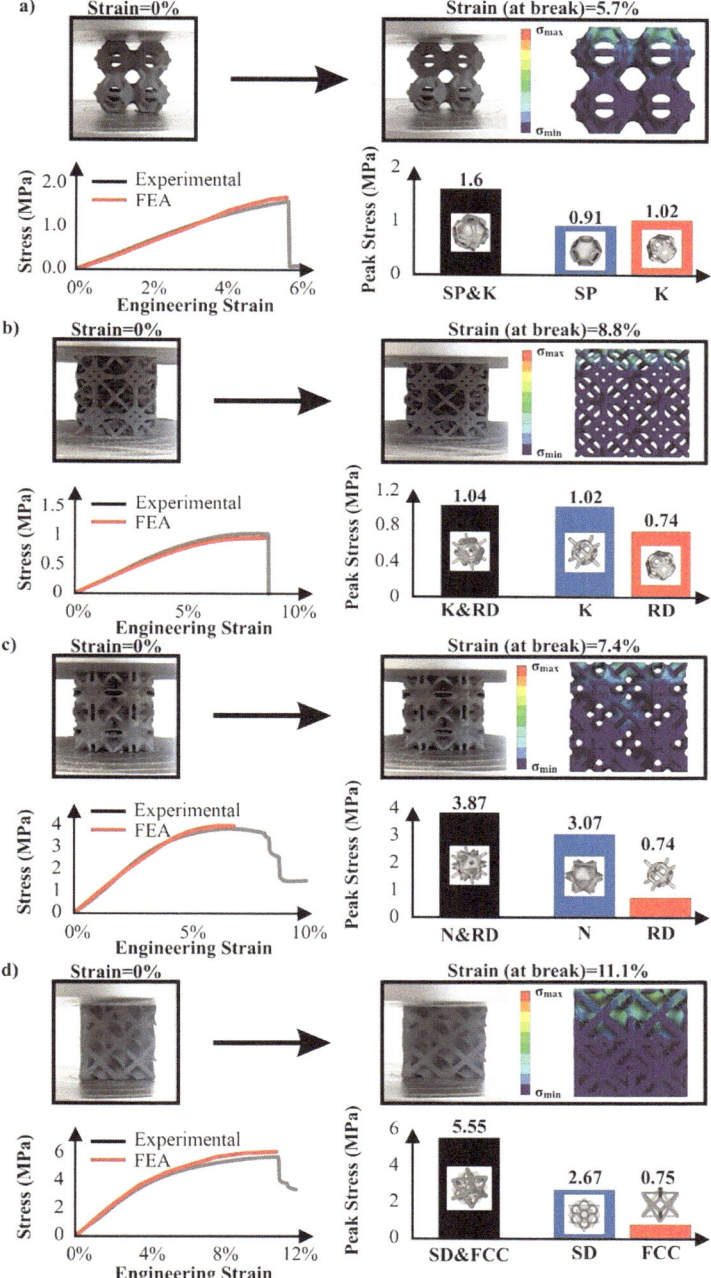

Fig. 5.8 Comparison of experimental data with FEA's results and origins mechanical performance for the four developed hybrid architected materials [43]

5.3.2 Functionality Gradation of Architected Materials

Functional gradation of an architected material refers to the intentional variation of effective properties within the same structure. This variation is achieved by modifying the applied relative density. The objective of this optimization process is the regulation of the relative density percentage in order to enhance the overall mechanical performance of the structure. Functional gradation is the most common and widespread optimization method for architected materials studied in many published scientific articles [36, 44–46]. Furthermore, the functional gradation process can be combined with an element-based TO approach to produce a topologically optimized structure with functional graded relative density of specific areas, where high stress concentration is observed. An indicative example of this procedure is presented in Fig. 5.9.

The functional gradation of the relative density for a structure consisting of architected material can be achieved through three distinct methods. The first method involves changing the thickness of the strut/walls of the architected material's element while keeping the length of the unit cell constant. The second method concerns the regulation of unit cell length by keeping the strut/wall thickness constant. Finally, the third method is a combination of the two aforementioned methods and results in final structures with greater geometric complexity. Figure 5.10 illustrates these three methods for the functional gradation of a simple structure consisting of 2 unit cells at 40% relative density using the sheet Gyroid architected material. As it is easily observed from the following example, the first produces a smoother transition from high to low relative density regions and extracts more uniform functionally

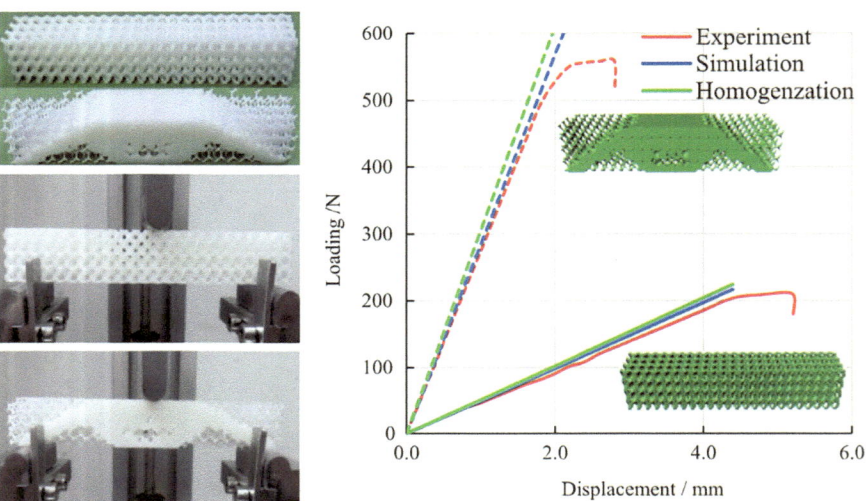

Fig. 5.9 Functional gradation combined with density-based TO approach for optimum bending performance [46] (Copyrights 2018 Elsevier)

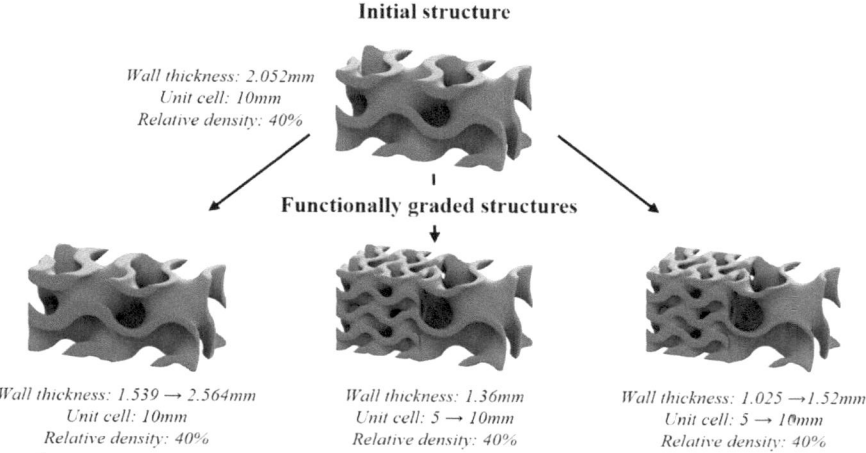

Fig. 5.10 Methods of functional gradation for the relative density of a lattice structure

graded structures. For these reasons, this method is utilized by the vast majority of the existing literature. It is worth noting that these three methods can be employed with four practices based on the rate of relative density modification. The first is the discrete practice with changes in relative density from unit cell to unit cell. The second and the third practices represent the linear and exponential ways where the relative density changes with linear and exponential dependence of a specific dimension, respectively. Finally, the last practice is the field-driven functional gradation of relative density, where the changes in relative density occur depending on a specific contour field.

Besides the methods of functional gradation of the structure's relative density, this optimization method can be utilized to induce other physical and mechanical properties on the structure. One of the most interesting applications of this method is the creation of lattice structures with low relative density on the external surface and a functional increase of the relative density towards the center of the structure, as it is depicted in the left side of Fig. 5.11. With this technique, the external region of the structure reveals high porosity and a rough external surface. Both of these can positively contribute to the improvement of the osseointegration process and the confrontation of wear effects in biomechanical applications, such as implants and scaffolds. On the other hand, functional gradation of the relative density can be employed to improve the mechanical and energy absorption performance of a structure by increasing the structure's relative density from the force's application region to the fixation, as portrayed in the right side of Fig. 5.11a. This method is very efficient in applications with high plastic strain, such as impact, crashworthiness, etc. [42].

Functionally graded architected materials have revealed interesting mechanical and energy absorption performance in applications with extensive plastic strain. More specifically, structures with top-to-bottom functional gradation of the relative

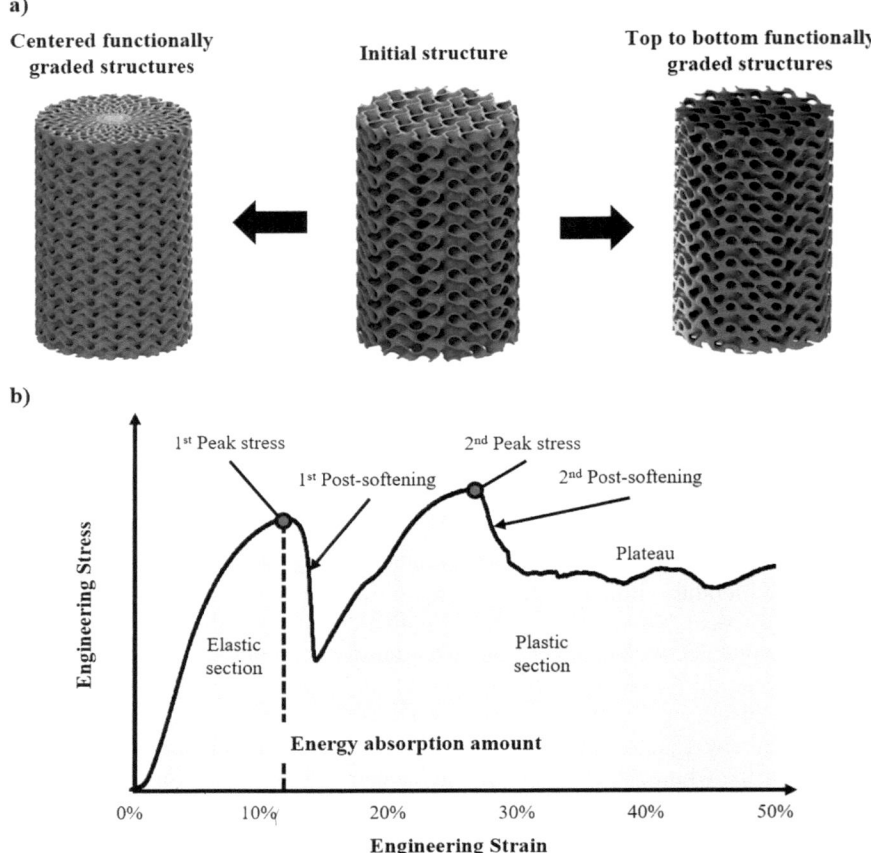

Fig. 5.11 a Indicative techniques of functional gradation to enhance physical and mechanical properties, **b** Indicative stress–strain for top-to-bottom functionally graded sheet Schwarz Diamond lattice structure

density exhibit high energy absorption capabilities, as it is illustrated in Fig. 5.11b. Figure 5.11b shows the unique stress–strain behavior of a functionally graded sheet SD structure with a 3 × 3 × 3 configuration and 30% relative density. It is worth mentioning that in this structure the functional gradation has been employed linearly. As mentioned previously, the structure starts to receive loads on the side with lower relative density and distributes them throughout its volume. Hence, during the elastic section, elastic strain is observed with increased stresses resulting in 1st peak stress. As the load increases, stresses are concentrated mainly on the contact region of the structure (i.e. low relative density region) resulting eventually in the breakage of the low-density unit cell. The destruction of the upper layer of unit cells leads to an extensively rapid drop in stresses, i.e. the 1st post-softening effect. However, the rest of the structure remains intact, and after the 1st post-softening effect, continues to receive loads, concentrating stresses on the unit cell layer with higher relative density. Then, the stress starts to increase again resulting in the 2nd peak stress. The 2nd peak stress is usually higher than the first due to the high mechanical strength of unit cells with increased relative density. After the 2nd peak stress, the loads exceed the strength of the unit cells, and their breakage is observed coupled with a 2nd post-softening effect that is smoother and milder. Because the examined structure has a $3 \times 3 \times 3$ configuration of unit cells, a plateau in increased stress is revealed after the 2nd post-softening effect. If the structure was composed of more unit cells, more peak stress points and post-softening effects would be observed. Nevertheless, after these phenomena, a plateau always occurs due to the high plastic strain compactness of the crashed structure. It is evident that all these peaks combined with the plateau, result in high energy absorption amounts.

In this context, the authors have performed an investigation utilizing the developed hybrid architected materials. In detail, functionally graded structures were developed with an ultimate goal, the exploitation of this optimization methodology in order to produce structures with increased strength and remarkable energy absorption. The examined architected materials were the Kelvin, the K&RD, the SD, and the SD&FCC. The functionally graded structures with bending-dominated behavior, i.e. Kelvin and K&RD, demonstrated lower stiffness and lower strength than the regular structures. This occurred due to the higher ductility and lower strength of the structures' struts with relatively small thicknesses. On the other hand, stretching-dominated structures, i.e. SD and SD&FCC, showed similar stiffness performance to their regular structure due to their comparable connectivity between the structure's unit cells. In addition, sufficiently higher peak strengths were observed for the stretching-dominated structures due to the stress distribution inside the structure's volume and the contribution to the overall performance of the unit cells with thicker wall/strut thicknesses. Figure 5.12 portrays indicative stress–strain diagrams for the examined functionally graded specimens at 30% total relative density along with superimposed images from the specimens at three testing phases, namely idle condition, 1st peak stress (1st breakage), and maximum peak stress. As it was expected, all the examined functionally graded structures had an increased surface area (energy absorption) below the stress–strain curve due to the existence of a plateau and the consecutive peak stresses. Furthermore, through the superimposed images, it is clearly visible that in the 1st peak stress all, the specimens concentrate the stress and the strain on the upper unit cells, which are in contact with the moving plate, resulting in local fracture without complete breakage of the specimens (the rest of the specimens seemed intact). However, during the highest peak stress loading, the specimens revealed stress and plastic strain throughout the structure's volume leading to complete breakage of the specimens and a rapid drop in stresses. It is worth mentioning that the bending-dominated structures (i.e. Kelvin and K&RD) mainly experienced compressive destruction during the maximum peak stress. In contrast, the stretching-dominated structures (i.e. SD and SD&FCC) exhibited both compressive and shear breakage of the specimens, with visible diagonal cracks. Overall, the functionally graded structures revealed mechanical behavior moved in the stretching-dominated realm compared to the mechanical behavior of the regular structures.

Table 5.6 lists the main effective mechanical properties of the aforementioned functionally graded structure revealing different performances depending on the mechanical behavior of each structure (stretching or bending dominated). In detail, the functionally graded structures with bending-dominated behavior, i.e. Kelvin and K&RD, demonstrated lower stiffness and lower strength than the regular structures. This occurred due to the higher ductility and lower strength of the structures' struts with relatively small thicknesses. On the other hand, stretching-dominated structures, i.e. SD and SD&FCC, showed similar stiffness performance to their regular structure due to their comparable connectivity between the structure's unit cells. In addition, significantly higher peak strengths were observed in the stretching-dominated structures due to the stress distribution inside the structure's volume and the contribution to the overall performance of the unit cells with thicker wall/strut thicknesses.

Moreover, Table 5.7 shows the SEA for these structures. It is obvious the energy absorption performance of all examined structures has been significantly improved in all indicators compared to their origins. Especially the stretching-dominated structures revealed a remarkable energy absorption amount with almost double the energy absorption amount of the architected materials with uniform relative density. Furthermore, it is worth noting that for SD & SD&FCC functionally graded structures, the calculated SEA values are greater than those of simple architected materials, i.e. 3D strut structures constructed with metallic materials. An indicative example of this is the comparison of SEA of 3D skeletal SD structure made of Maraging steel with density at 10^{-3} g/mm^3. More specifically, according to the existing literature [47], this structure for this density revealed a SEA of 1.2 J/g in contrast to the corresponding PA12 functionally graded specimens, which showed SEA values close to 2 J/g. It is assumed that this difference was observed due to the severe post-softening effect of the metal structure and the absence of a plateau with sufficient stresses.

5.3.3   Interpenetrating Phase Composite Structures

Interpenetrating Phase Composite (IPC) is an optimization method for architected material utilized to enhance the physical and mechanical properties of a structure. IPCs are created by filling the void space of an architected material structure with filling materials (2nd phase) [48]. These structures are composites with the unique characteristic of interconnectivity between the combined phases. The interconnectivity is so intense that even if one phase is completely removed the other phase remains intact, forming a self-supported structure. The physical and mechanical properties of the IPCs are mostly influenced by the spatial distribution of the material's phase and their volume percentage in the overall structure. Thus, the properties of these structures can be easily manipulated by controlling the spatial phase distribution. IPC can be fabricated employing several manufacturing methods, such as multi-material 3D printing. Moreover, another fabrication technique involves filling a

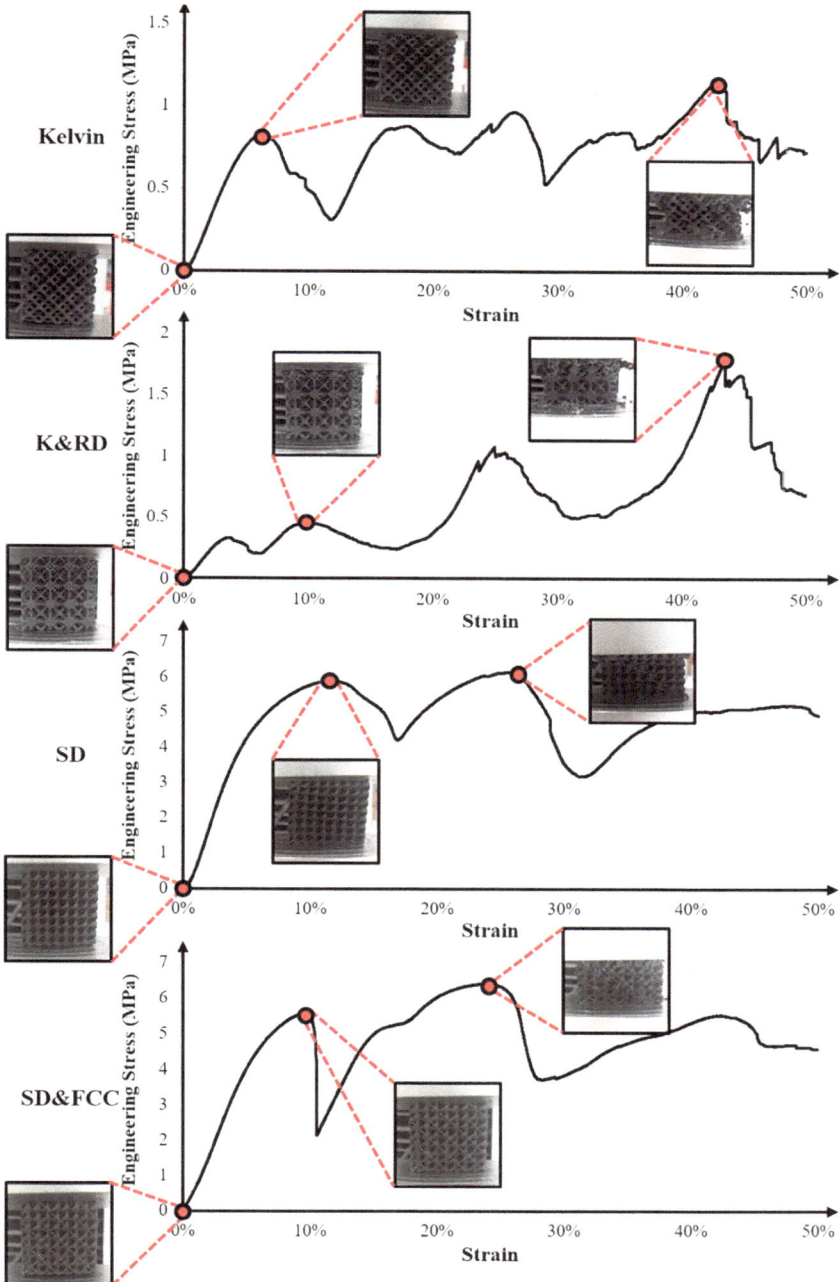

Fig. 5.12 Indicative stress–strain diagrams for the examined functionally graded structure at 30% total relative density coupled with superimposed images of the fracture modes

Table 5.6 Effective properties for the examined functionally graded architected materials

Functionally graded architected materials	Effective elastic modulus			Effective peak stress		
	20%	30%	40%	20%	30%	40%
Kelvin	3.38	17.57	33.2	0.473	1.151	1.786
K&RD	9.38	14.29	28.57	0.474	1.777	2.877
SD	99.87	108.33	115.68	5.949	6.245	7.077
SD&FCC	84.9	96.49	114.04	6.079	6.394	7.192

Table 5.7 SEA of all examined functionally graded architected materials for all examined relative density

Functionally graded architected materials	SEA (J/g)		
	Relative densities		
	20%	30%	40%
Kelvin	0.128	0.387	0.568
K&RD	0.134	0.340	0.642
SD	2.078	2.493	2.741
SD&FCC	2.362	2.448	2.828

solid architected material structure with a liquid material that will be solidified during the process, such as silicon rubber compounds or construction foams, etc. There are plenty of studies that investigated the creation of IPCs, both numerically and experimentally, in order to develop a final structure with enhanced thermal, electrical, and mechanical properties [18, 49–51]. More specifically, the study of Abueidda et al. [49] on which IPC structures, utilizing sheet TPMS architected, were developed with enhanced effective thermal conductivity and low thermal expansion compared to other structural composites, like fiber-reinforced and sheet-reinforced composites. Furthermore, IPCs with architected materials were fabricated bio-mimicking the structure of natural bones with a hard shell and a soft core. In this study, Al-Ketan et al. [50] examined and evaluated the mechanical properties of IPC structures highlighting the increased toughness of these structures and showing their high potential in terms of energy absorption performance. Finally, IPCs with architected materials have been proposed for several other engineering applications due to their enhanced recoverability. An indicative example is their employment on joints of a tendon-driven soft robotic finger [51]. In this application, a rigid material was used for the construction of IPC's sheet TPMS core and an elastomer material was employed as the matrix of the IPC structure. In this way, the soft matrix offered the necessary recoverability of the joint, while the hard core controlled the deformation level of the joint.

Based on the literature review, it is evident that the optimization method of IPCs holds significant potential for energy absorption applications. Therefore, in this context, IPC structures with functionally graded architected materials were developed, for the purposes of this book, by exploiting the acquired results of previous analyses in order to produce structures with increased strength and remarkable energy absorption. These IPC structures were selected to be sandwich-like structures. Overall, sandwich-like structures offer many benefits, including a high strength-to-weight ratio, increased stiffness, improved insulation, vibration damping, and high energy absorption. Hence, these structures are used in a wide range of applications, including aerospace, automotive, construction, and sports equipment. Therefore, the next step was the creation of sandwich-like structures employing the most promising architected materials from the previous evaluation for the functionally graded structures, i.e. the SD and SD&FCC structures. The sandwich-like structures were developed with the addition of solid/bulk layers to the upper and lower sides of the functionally graded structures. These layers are named as facings and had a thickness of 10% of the overall height of the functionally graded specimens, hence the facing thickness was selected at 3 mm. Then, the developed sandwich-like structures were utilized as a rigid phase for the creation of IPC. In other words, the void spaces of the developed sandwich-like structures were filled with foam materials (2nd phase) which solidified after a time period. As foam material, a commonly used material in construction sides was used, namely polyurethane construction foam, due to three reasons. First, polyurethane construction foam could be easily formed after the extraction from the bottle and fitted in the 1st phase material. Secondly, after the solidification of the material, polyurethane becomes a rigid foam with varying relative density (from 10 to 50%) and close to stretching-dominated behavior ($n_E \approx$ 1.6). The approximate values of elastic modulus and yield strength range from 50 to 300 MPa and from 1 to 15 MPa, respectively. These remarkable properties of polymer foam with stochastic structure make this material one of the best candidates for this application. Finally, polyurethane material exhibits intense adhesive behavior which is essential for the combination of materials in IPCs. The IPCs were fabricated with the employment of special molds and highly pressurized polyurethane foam with a low expansion rate. In detail, the molds were designed to properly fit the sandwich-like structure and hold it during the pressurization process. Also, the developed molds had multiple injection holes for the insertion of the polyurethane foam from different positions to ensure the complete filling of the structure. The designed molds were fabricated utilizing the Fused Filament Fabrication (FFF) 3D printing technique. Then, the sandwich-like structures were assembled and secured in a fixed position with the mold. The next process was the insertion of the polyurethane foam. The polyurethane foam was in a compressed bottle of 13bar and with the assistance of a nozzle was injected into the desired positions. With this method, a complete filling of the sandwich-like structures was observed. Figure 5.13 portrays an indicative image of an IPC sandwich-like structure constructed with SD architected material and polyurethane foam. It is worth noting that the fabrication process is critical in order to produce an object with relatively uniform foam distribution and optimize the overall procedure as incomplete areas i.e. without polyurethane foam, might be present as illustrated in the current case.

Fig. 5.13 Indicative IPC
sandwich-like structure with
SD architected material and
polyurethane foam

The evaluation of the mechanical behavior for the developed IPC sandwich-like structures was performed with the same methodology of the selected functionally graded architected materials, i.e. SD and SD&FCC. In detail, specimens with three different total relative densities of 20%, 30%, and 40% were designed, fabricated, and tested to extract the values of the effective mechanical properties, calculate the constant of the applied scaling laws, and observe the fracture modes for these new structures.

Tables 5.8 and 5.9 present the values of the effective mechanical properties and the values of the constant of the scaling laws, respectively, for the examined IPC functionally graded sandwich-like structures. According to the acquired experimental data of the effective elastic modulus the IPC sandwich-like structure revealed higher stiffness for all relative densities with an increase of more than 10%. Furthermore, a similar pattern was observed for the effective peak strength with a higher increase at higher relative densities. Both phenomena were expected and mainly occurred due to the superior connectivity provided by the polyurethane foam and the enhancement of structural integrity caused by the addition of the upper and lower plates. Moreover, these elements led to a more intense stretching-dominated behavior, as it is clearly indicated by the values of n_E in Table 5.9. It is worth mentioning that the almost logarithmic relation of the effective mechanical properties with the relative density mainly occurred on the new micro-porosity of the structure caused by the polyurethane foam.

Table 5.8 Effective properties for the examined IPC functionally graded sandwich-like structures

IPC functionally graded sandwich-like structure	Effective elastic modulus			Effective peak stress		
	20%	30%	40%	20%	30%	40%
SD	109.12	123.99	136.26	6.125	6.586	7.007
SD&FCC	111.24	124.98	145.33	6.111	7.275	8.174

Table 5.9 Values for the constants of the scaling laws

IPC functionally graded sandwich-like structure	Effective elastic modulus		Effective peak stress	
	G_E	n_E	$G_{\sigma p}$	$n_{\sigma p}$
SD	0.19	0.32	0.26	0.19
SD&FCC	0.38	0.42	0.22	0.39

Figure 5.14 illustrates indicative stress–strain diagrams for the examined IPC sandwich-like structure at 40% total relative density coupled with superimposed images of the fracture modes. Comparing these diagrams with the corresponding diagrams of the previous subchapter, the following points are observed. Firstly, the consecutive peaks in the stress–strain diagram have been minimized with the highest peak occurring right after the yield point. Secondly, the IPC structures withstood greater loads than the simple functionally graded structures. Thirdly, due to the existence of the polyurethane foam, the rapid drops of stresses, known as postsoftening effects, have been smoothened and the drops are relatively smaller. Finally, the existing plateau of the IPC structures occurred at higher stresses than the other structure resulting in high energy absorption. The superimposed images of the structure in crucial conditions during the compressive loading showed that the stresses are distributed more uniformly inside the structure's volume and even after extensive plastic strain, the structures remained compact and maintained their structural integrity.

Regarding the energy absorption of the IPC functionally graded sandwich-like structure, it was measured again with an identical methodology as in the previous subchapter in order to extract comparable results. Table 5.10 lists the SEA value for these structures for each relative density. As expected from the analysis of the stress–strain diagram, the energy absorption performance was significantly improved especially for the structure with the SD architected material. More specifically, the IPC functionally graded sandwich-like structures consisting of SD architected material experienced an increase of SEA with a mean value of 14.3%, whereas the structures consisting of SD&FCC architected material exhibited a slightly lower mean increase of SEA at 13.1%. Another important observation of the IPC functionally graded sandwich-like structures is that their energy efficiency remained practically the same as that of the functionally graded structure. This indicates that the IPC functionally graded sandwich-like structure had higher peak stress (which is verified), but lower mean stresses through the stress–strain diagram. Of course, this response can be explained by the multiple consecutive peak stress that occurred in pure functionally graded structures, resulting in a higher mean stress value compared with their highest stress.

To conclude, in the context of this chapter, three optimization methods for architected materials were presented and examined through a series of experiments. More specifically, by utilizing hybrid architected materials, the relative density's functional

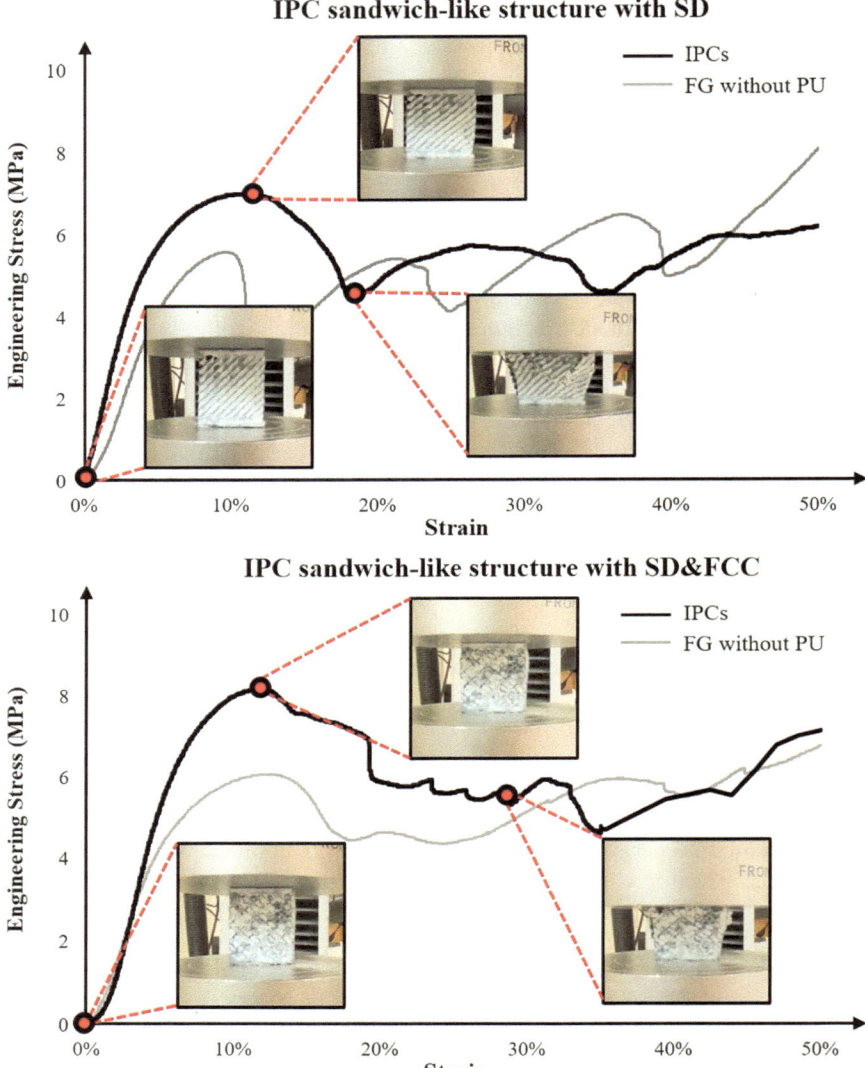

Fig. 5.14 Indicative stress–strain diagrams for the examined IPC sandwich-like structure at 40% total relative density coupled with superimposed images of the fracture modes

gradation and the fabrication of IPCs, the production of sandwich-like panel structures with almost three times higher SEA values, was achieved, as it is presented in Table 5.11. These novel structures can be implemented in several applications with high demands in energy absorption capabilities, crashworthiness, etc.

Table 5.10 SEA of all examined IPC functionally graded sandwich-like structures for all examined relative densities

IPC functionally graded sandwich-like structure	SEA (J/g)		
	Relative densities		
	20%	30%	40%
SD	2.486	2.858	3.176
SD&FCC	2.488	2.962	3.405

Table 5.11 Comparison table of the SEA improvement for the examined structures at 20% relative density

Architected material	SEA of regular lattice structures (J/g)	SEA of developed sandwich-like structures (J/g)	Percentage increase (%)
SD	0.951	2.858	201
SD&FCC	1.028	2.962	188

References

1. N. Kladovasilakis, K. Tsongas, I. Kostavelis, D. Tzovaras, D. Tzetzis, Effective mechanical properties of additive manufactured triply periodic minimal surfaces: experimental and finite element study. Int. J. Adv. Manuf. Technol. **121**(11–12), 7169–7189 (2022). https://doi.org/10.1007/s00170-022-09651-w

2. I. Maskery, L. Sturm, A.O. Aremu, A. Panesar, C.B. Williams, C.J. Tuck, R.D. Wildman, I.A. Ashcroft, R.J.M. Hague, Insights into the mechanical properties of several triply periodic minimal surface lattice structures made by polymer additive manufacturing. Polymer (Guildf) **152**, 62–71 (2018). https://doi.org/10.1016/j.polymer.2017.11.049

3. M. Benedetti, A. du Plessis, R.O. Ritchie, M. Dallago, S.M.J. Razavi, F. Berto, Architected cellular materials: a review on their mechanical properties to-wards fatigue-tolerant design and fabrication. Mater. Sci. Eng. R. Rep. **144**, 100606 (2021). https://doi.org/10.1016/j.mser.2021.100606

4. C. Yan, L. Hao, A. Hussein, P. Young, D. Raymont, Advanced lightweight 316L stainless steel cellular lattice structures fabricated via selective laser melting. Mater. Des. **55**, 533–541 (2014). https://doi.org/10.1016/j.matdes.2013.10.027

5. F.S.L. Bobbert, K. Lietaert, A.A. Eftekhari, B. Pouran, S.M. Ahmadi, H. Weinans, A.A. Zadpoor, Additively manufactured metallic porous bio-materials based on minimal surfaces: a unique combination of topological, mechanical, and mass transport properties. Acta Biomater. **53**, 572–584 (2017). https://doi.org/10.1016/j.actbio.2017.02.024

6. J. Kadkhodapour, H. Montazerian, A.C. Darabi, A.P. Anaraki, S.M. Ahmadi, A.A. Zadpoor, S. Schmauder, Failure mechanisms of additively manufactured porous biomaterials: effects of porosity and type of unit cell. J. Mech. Behav. Biomed. Mater. **50**, 180–191 (2015). https://doi.org/10.1016/j.jmbbm.2015.06.012

7. S. Ahmadi, S. Yavari, R. Wauthle, B. Pouran, J. Schrooten, H. Weinans, A. Zadpoor, Additively manufactured open-cell porous biomaterials made from six different space-filling unit cells: the mechanical and morphological properties. Materials **8**(4), 1871–1896 (2015). https://doi.org/10.3390/ma8041871

8. N. Novak, O. Al-Ketan, M. Borovinšek, L. Krstulović-Opara, R. Rowshan, M. Vesenjak, Z. Ren, Development of novel hybrid TPMS cellular lattices and their mechanical characterisation. J. Market. Res. **15**, 1318–1329 (2021). https://doi.org/10.1016/j.jmrt.2021.08.092

9. O. Al-Ketan, R. Rowshan, R.K. Abu Al-Rub, Topology-mechanical property relationship of 3D printed strut, skeletal, and sheet based periodic metallic cellular materials. Addit. Manuf. **19**, 167–183 (2018). https://doi.org/10.1016/j.addma.2017.12.006

10. F.S.L. Bobbert, K. Lietaert, A.A. Eftekhari, B. Pouran, S.M. Ahmadi, H. Weinans, A.A. Zadpoor, Additively manufactured metallic porous bio-materials based on minimal surfaces: a unique combination of topological, mechanical, and mass transport properties. Acta Biomater. **53**, 572–584 (2017). https://doi.org/10.1016/j.actbio.2017.02.024

11. L. Zhang, S. Feih, S. Daynes, S. Chang, M.Y. Wang, J. Wei, W.F. Lu, Energy absorption characteristics of metallic triply periodic minimal surface sheet structures under compressive loading. Addit. Manuf. **23**, 505–515 (2018). https://doi.org/10.1016/j.addma.2018.08.007

12. D.W. Abueidda, M. Bakir, R.K. Abu Al-Rub, J.S. Bergström, N.A. Sobh, I. Jasiuk, Mechanical properties of 3D printed polymeric cellular materials with triply periodic minimal surface architectures. Mater. Des. **122**, 255–267 (2017). https://doi.org/10.1016/j.matdes.2017.03.018

13. O. Al-Ketan, R.K. Abu Al-Rub, R. Rowshan, The effect of architecture on the mechanical properties of cellular structures based on the IWP minimal surface. J. Mater. Res. **33**(3), 343–359 (2018). https://doi.org/10.1557/jmr.2018.1

14. C. Yan, L. Hao, A. Hussein, S.L. Bubb, P. Young, D. Raymont, Evaluation of light-weight AlSi10Mg periodic cellular lattice structures fabricated via direct metal laser sintering. J. Mater. Process. Technol. **214**(4), 856–864 (2014). https://doi.org/10.1016/j.jmatprotec.2013.12.004

15. A.M. Abou-Ali, O. Al-Ketan, R. Rowshan, R. Abu Al-Rub, Mechanical response of 3D printed bending-dominated ligament-based triply periodic cellular polymeric solids. J. Mater. Eng. Perform. **28**(4), 2316–2326 (2019). https://doi.org/10.1007/s11665-019-03982-8

16. N. Kladovasilakis, K. Tsongas, I. Kostavelis, D. Tzovaras, D. Tzetzis, Effective mechanical properties of additive manufactured strut-lattice structures: experimental and finite element study. Adv. Eng. Mater. **24**(3), 2100879 (2022). https://doi.org/10.1002/adem.202100879

17. H. Jiao, Q. Zhou, S. Fan, Y. Li, A new hybrid topology optimization method coupling ESO and SIMP method (2015) pp. 373–384. https://doi.org/10.1007/978-3-662-44674-4_35

18. O. Al-Ketan, R.K. Abu Al-Rub, Multifunctional mechanical metamaterials based on triply periodic minimal surface lattices. Adv. Eng. Mater. **21**(10), 1900524 (2019). https://doi.org/10.1002/adem.201900524

19. N. Kladovasilakis, K. Tsongas, D. Karalekas, D. Tzetzis, Architected materials for additive manufacturing: a comprehensive review. Materials **15**(17), 5919 (2022). https://doi.org/10.3390/ma15175919

20. S. Al Hassanieh, A. Alhantoobi, K.A. Khan, M.A. Khan, Mechanical properties and energy absorption characteristics of additively manufactured light-weight novel re-entrant plate-based lattice structures. Polymers (Basel) **13**(22), 3882 (2021). https://doi.org/10.3390/polym13223882

21. T. Femmer, A.J.C. Kuehne, M. Wessling, Estimation of the structure dependent performance of 3-D rapid prototyped membranes. Chem. Eng. J. **273**, 438–445 (2015). https://doi.org/10.1016/j.cej.2015.03.029

22. O. Al-Ketan, M. Pelanconi, A. Ortona, R.K. Abu Al-Rub, Additive manufacturing of architected catalytic ceramic substrates based on triply periodic minimal surfaces. J. Am. Ceram. Soc. **102**(10), 6176–6193 (2019). https://doi.org/10.1111/jace.16474

23. D.W. Abueidda, R.K. Abu Al-Rub, A.S. Dalaq, D.-W. Lee, K.A. Khan, I. Jasiuk, Effective conductivities and elastic moduli of novel foams with triply periodic minimal surfaces. Mech. Mater. **95**, 102–115 (2016). https://doi.org/10.1016/j.mechmat.2016.01.004

24. I. Maskery, A.O. Aremu, L. Parry, R.D. Wildman, C.J. Tuck, I.A. Ashcroft, Effective design and simulation of surface-based lattice structures featuring volume fraction and cell type grading. Mater. Des. **155**, 220–232 (2018). https://doi.org/10.1016/j.matdes.2018.05.058

25. S.C. Kapfer, S.T. Hyde, K. Mecke, C.H. Arns, G.E. Schröder-Turk, Minimal surface scaffold designs for tissue engineering. Biomaterials **32**(29), 6875–6882 (2011). https://doi.org/10.1016/j.biomaterials.2011.06.012

26. A. Yáñez, A. Cuadrado, O. Martel, H. Afonso, D. Monopoli, Gyroid porous titanium structures: a versatile solution to be used as scaffolds in bone defect reconstruction. Mater. Des. **140**, 21–29 (2018). https://doi.org/10.1016/j.matdes.2017.11.050

27. J. Kadkhodapour, H. Montazerian, S. Raeisi, Investigating internal architecture effect in plastic deformation and failure for TPMS-based scaffolds using simulation methods and experimental procedure. Mater. Sci. Eng. C **43**, 587–597 (2014). https://doi.org/10.1016/j.msec.2014.07.047

28. C. Bonatti, D. Mohr, Smooth-shell metamaterials of cubic symmetry: anisotropic elasticity, yield strength and specific energy absorption. Acta Mater. **164**, 301–321 (2019). https://doi.org/10.1016/j.actamat.2018.10.034

29. L. Yang, C. Yan, H. Fan, Z. Li, C. Cai, P. Chen, Y. Shi, S. Yang, Investigation on the orientation dependence of elastic response in gyroid cellular structures. J. Mech. Behav. Biomed. Mater. **90**, 73–85 (2019). https://doi.org/10.1016/j.jmbbm.2018.09.042

30. K.A. Khan, R.K. Abu Al-Rub, Time dependent response of architectured neovius foams. Int. J. Mech. Sci. **126**, 106–119 (2017). https://doi.org/10.1016/j.ijmecsci.2017.03.017

31. J.-H. Park, J.-C. Lee, Unusually high ratio of shear modulus to young's modulus in a nano-structured gyroid metamaterial. Sci. Rep. **7**(1), 10533 (2017). https://doi.org/10.1038/s41598-017-10978-8

32. G.S. Jung, J. Yeo, Z. Tian, Z. Qin, M.J. Buehler, Unusually low and density-insensitive thermal conductivity of three-dimensional gyroid graphene. Nanoscale **9**(36), 13477–13484 (2017). https://doi.org/10.1039/C7NR04455K

33. M. Zhao, F. Liu, G. Fu, D. Zhang, T. Zhang, H. Zhou, Improved mechanical properties and energy absorption of BCC lattice structures with triply periodic minimal surfaces fabricated by SLM. Materials **11**(12), 2411 (2018). https://doi.org/10.3390/ma11122411

34. Y.L.A. Alshammari, F. He, M.A. Khan, Modelling and investigation of crack growth for 3D-printed acrylonitrile butadiene styrene (ABS) with various printing parameters and ambient temperatures. Polymers (Basel) **13**(21), 3737 (2021). https://doi.org/10.3390/polym13213737

35. Y. Liu, Mechanical properties of a new type of plate-lattice structures. Int. J. Mech. Sci. **192**, 106141 (2021). https://doi.org/10.1016/j.ijmecsci.2020.106141

36. L. Yang, R. Mertens, M. Ferrucci, C. Yan, Y. Shi, S. Yang, Continuous graded gyroid cellular structures fabricated by selective laser melting: design. Manufact. Mech. Prop. Mater Des. **162**, 394–404 (2019). https://doi.org/10.1016/j.matdes.2018.12.007

37. L. Zhang, S. Feih, S. Daynes, S. Chang, M.Y. Wang, J. Wei, W.F. Lu, Energy absorption characteristics of metallic triply periodic minimal surface sheet structures under compressive loading. Addit. Manuf. **23**, 505–515 (2018). https://doi.org/10.1016/j.addma.2018.08.007

38. T. Tancogne-Dejean, M. Diamantopoulou, M.B. Gorji, C. Bonatti, D. Mohr, 3D plate-lattices: an emerging class of low-density metamaterial exhibiting optimal isotropic stiffness. Adv. Mater. **30**(45), 1803334 (2018). https://doi.org/10.1002/adma.201803334

39. A. Baroutaji, M. Sajjia, A.-G. Olabi, On the crashworthiness performance of thin-walled energy absorbers: recent advances and future developments. Thin-Walled Struct. **118**, 137–163 (2017). https://doi.org/10.1016/j.tws.2017.05.018

40. J. Fang, G. Sun, N. Qiu, N.H. Kim, Q. Li, On design optimization for structural crashworthiness and its state of the art. Struct. Multidiscip. Optim. **55**(3), 1091–1119 (2017). https://doi.org/10.1007/s00158-016-1579-y

41. C.R. Siviour, *High Strain Rate Characterization of Polymers* (2017), p. 060029. https://doi.org/10.1063/1.4971585

42. N. Novak, O. Al-Ketan, L. Krstulović-Opara, R. Rowshan, R.K. Abu Al-Rub, M. Vesenjak, Z. Ren, Quasi-static and dynamic compressive behaviour of sheet TPMS cellular structures. Compos. Struct. **266**, 113801 (2021). https://doi.org/10.1016/j.compstruct.2021.113801

43. N. Kladovasilakis, K. Tsongas, D. Tzetzis, Development of novel additive manufactured hybrid architected materials and investigation of their mechanical behavior. Mech. Mater. 176 (2023). https://doi.org/10.1016/j.mechmat.2022.104525

44. A. Panesar, M. Abdi, D. Hickman, I. Ashcroft, Strategies for functionally graded lattice structures derived using topology optimisation for additive manufacturing. Addit. Manuf. **19**, 81–94 (2018). https://doi.org/10.1016/j.addma.2017.11.008

45. M. Afshar, A. Pourkamali Anaraki, H. Montazerian, Compressive characteristics of radially graded porosity scaffolds architectured with minimal surfaces. Mater. Sci. Eng. C **92**, 254–267 (2018). https://doi.org/10.1016/j.msec.2018.06.051

46. D. Li, W. Liao, N. Dai, G. Dong, Y. Tang, Y.M. Xie, Optimal design and modeling of gyroid-based functionally graded cellular structures for additive manufacturing. Comput. Aided Des. **104**, 87–99 (2018). https://doi.org/10.1016/j.cad.2018.06.003

47. Berhanu, S.; Tariq, F.; Jones, T.; McComb, D. W. Three-Dimensionally Interconnected Organic Nanocomposite Thin Films: Implications for Donor–Acceptor Photovoltaic Applications. J Mater Chem 2010, 20 (37), 8005. https://doi.org/10.1039/c0jm01030h

48. Z. Poniznik, V. Salit, M. Basista, D. Gross, Effective elastic properties of interpenetrating phase composites. Comput. Mater. Sci. **44**(2), 813–820 (2008). https://doi.org/10.1016/j.commatsci.2008.06.010

49. D.W. Abueidda, R.K. Abu Al-Rub, A.S. Dalaq, H.A. Younes, A.A. Al Ghaferi, T.K. Shah, Electrical conductivity of 3D periodic architectured interpenetrating phase composites with carbon nanostructured-epoxy reinforcements. Compos. Sci. Technol. **118**, 127–134 (2015). https://doi.org/10.1016/j.compscitech.2015.08.021

50. O. Al-Ketan, A. Soliman, A.M. AlQubaisi, R.K. Abu Al-Rub, Nature-inspired lightweight cellular co-continuous composites with architected periodic gyroidal structures. Adv. Eng. Mater. **20**(2), 1700549 (2018). https://doi.org/10.1002/adem.201700549

51. I. Hussain, O. Al-Ketan, F. Renda, M. Malvezzi, D. Prattichizzo, L. Seneviratne, R.K. Abu Al-Rub, D. Gan, Design and prototyping soft-rigid tendon-driven modular grippers using interpenetrating phase composites materials. Int. J. Rob. Res. **39**(14), 1635–1646 (2020). https://doi.org/10.1177/0278364920907697

Chapter 6
Case Studies of Topology Optimization

In the current chapter, developed case studies are presented and analyzed utilizing both the element-based approach and the discrete/truss-based approach with the utilization of architected materials. The aim of this chapter is to demonstrate the broad applicability and functionality of topology optimization processes across various industries, including automotive, construction, and bioengineering.

6.1 Element-Based Approach Applications

The element-based method is extensively used in a plethora of applications, mostly through the utilization of the SIMP algorithm. Topologically optimized components are necessary for several fields, such as mechanical and biomechanical engineering, robotics, automotive, and aerospace, which require lightweight components with optimum mechanical performance. In the context of this book, three individual preliminary case studies were performed in order to topologically optimize a bone implant (i.e. tibial implant for a knee replacement), an automotive brake caliper, and a lower prosthetic limb. All these studies utilized density-based TO process with SIMP algorithm. Furthermore, the initial and final designs are presented in Fig. 6.1 along with the stress concentration contours of the initial designs. In addition, through the density-based TO process mass reduction was achieved, ranging from 24.8% to 67% for the examined case studies without compromising the structural integrity of the structure, i.e. the maximum stress was below the material's yield strength. Table 6.1 lists the initial and final mass, as well as the mass reduction percentage before and after the density-based TO of the three examined components. It is pertinent to note that the mass reduction percentage is highly varied in each case, depending on the construction material, the applied loads, the fidelity of the employed TO process, and the bulkiness of the initial design [1, 2, 3].

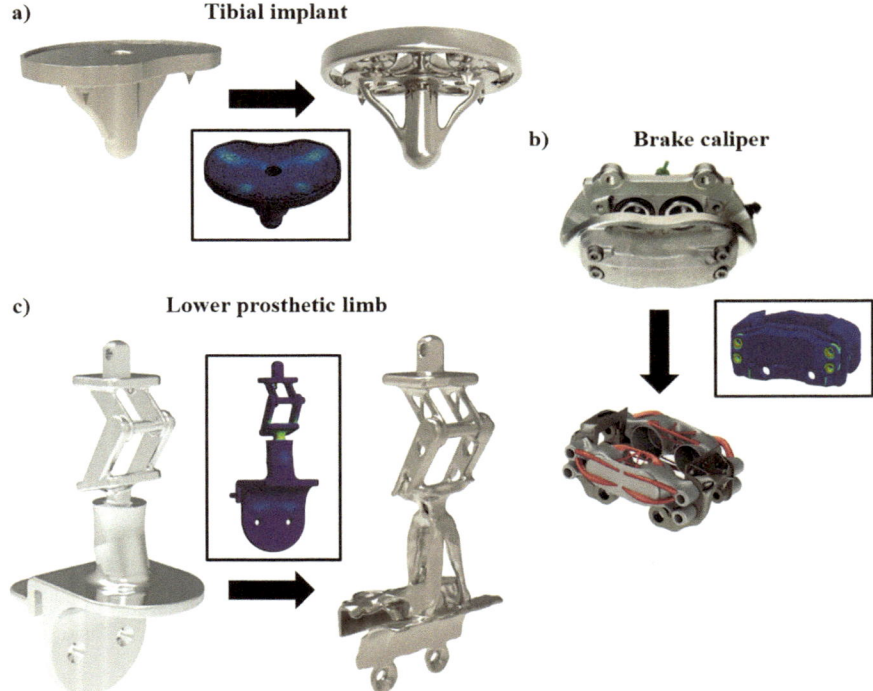

Fig. 6.1 Case studies of the density-based TO with SIMP algorithm: **a** tibial implant [1], **b** automotive brake caliper [2], and c) lower prosthetic limb [3]

Table 6.1 Initial and final mass coupled with the mass reduction percentage before and after density-based TO of the three individual components

Components	Initial mass (g)	Final mass (g)	Mass reduction (%)
Tibial implant	177.69	133.68	24.8
Brake caliper	3195	2173	32
Prosthetic limb	3460	1130	67

Topology optimization is emerging as one of the most promising design methods, especially in applications where weight is a critical factor. A notable instance of this trend was the 2014 open challenge by General Electric (GE), which invited designers worldwide to topologically optimize a commercial jet engine bracket [4]. Figure 6.2 shows the results of the topology optimization process from a relevant study on this bracket [5]. In this study, a 65% weight reduction was achieved, from 2.067 kg (initial design in Fig. 6.2a) to 0.720 kg final design in Fig. 6.2b. Moreover, the final design could withstand the operation loads, revealing a sufficient safety factor of 1.11. Similar to this case study, many enterprises worldwide have commercially used the topology optimization processes of element-based approach for their parts, especially

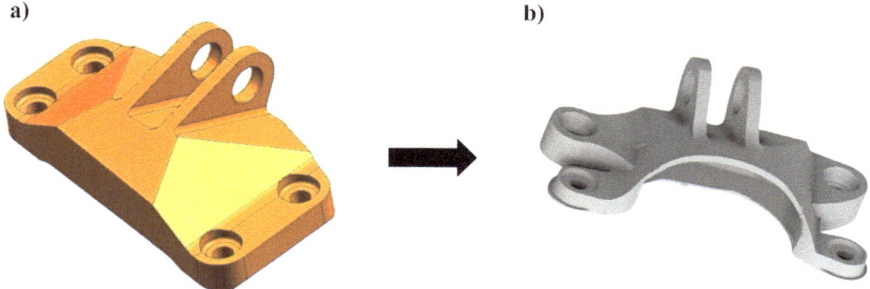

Fig. 6.2 a Initial design, **b** topologically optimized version of the commercial jet engine bracket
[5]

in the automotive and aeronautical industries. Through this process, lightweight
(high stiffness-to-weight ratio) and in many cases aesthetically enhanced commercial
components are produced.

6.2 Discrete/Truss-Based Approach Applications

As previously mentioned, the discrete/truss-based approach for topology optimization is highly complex, requiring extensive technical knowledge and research. Therefore, its commercial adoption has encountered obstacles, prompting global efforts
to move beyond the proof-of-concept stage and achieve practical applicability in
real-world scenarios. The increased interest in this topology optimization approach
is derived from the competitive advantages that offers in the produced part. Besides
the lightweight of the final structures, this approach also can offer tailored mechanical properties through the manipulation of relative density, high porosity, and high
surface-to-volume ratio that are significant in a wide range of applications, such as
bioengineering, heat transfer, sound absorption, etc. In this context, the book presents
indicative examples of architected materials utilized in several applications based on
studies that were conducted during the preparation of this book.

6.2.1 2.5D Architected Materials Applications

2.5D architected materials are the simplest form of architected materials that can be
used to enhance and optimize an existing structure. In this subchapter, three different
case studies are presented that have already been published. The first case study
focuses on the topology optimization of an air manifold [6], the second examines the
performance enhancement of elevator braking pads through topology optimization

[7], and the third addresses the improvement of reinforcement structures for clay-based constructions [8].

Air-manifold:

In the current work, a workflow for redesigning and topologically optimizing a low-pressure air manifold was developed and used with the aid of cutting-edge design processes and Selective Laser Sintering (SLS) 3D printing with polyamide 12 (PA12). As the first step, the element-based topology optimization approach was utilized, via SIMP algorithm, in order to remove unnecessary material regions and then conformal 2.5D surface-architected materials were applied to critical areas to enhance its structural integrity. The final design assessed computationally under realistic conditions, achieved a weight reduction of up to 76% compared to the original design, with production costs under €5 per unit. The workflow presented offers a generic pipeline for the redesign and topology optimization of a manifold, which can be applied to other similar 3D-printed mechanical components made from either metal or polymer, such as hydraulic manifolds, nozzles, etc. Table 6.2 lists the volume, mass and factor of safety (FOS) for the initial and developed designs. It is obvious that the element-based approach was not enough for this kind of application resulting in insufficient FOS value. However, the integration of a conformal lattice structure with 2.5D surface-architected material configuration provided necessary structural rids on the structure increasing its mechanical strength and leading to a topologically optimized final design. Moreover, Fig. 6.3(a–d) summarizes the main points of this study indicating both the corresponding methodology, 3D models and numerical evaluation. The same methodology has found increased applicability in aerospace and automotive, especially in pressurized components, such as engine blocks and Laval nozzles.

Elevator braking pads:

In this case study novel bioinspired structures for elevator safety gear friction pads were 3D designed, 3D printed, and evaluated with the aim of improving the pads' dynamic friction response and minimizing undesired effects observed in conventional pads [7]. Four different friction pads integrated with surface lattice structures of 2.5D bioinspired architected materials were employed on the friction surface of the conventional pads and 3D printed by selective laser melting (SLM) process utilizing tool steel H13 powder as construction material. Each pad was examined with tribological tests to assess its dynamic coefficient of friction (CoF). Based on the acquired CoFs, finite element models (FEM) were developed to evaluate their

Table 6.2 Volume, mass, and factor of safety values for all examined versions of the air manifold

Designs	Volume (cm^3)	Mass (g)	Difference	FOS
Initial version	80.5	78.09	–	1.02
Optimized version	17.5	16.98	78.3%	0.76
Final version	19.7	19.11	75.5%	1.04

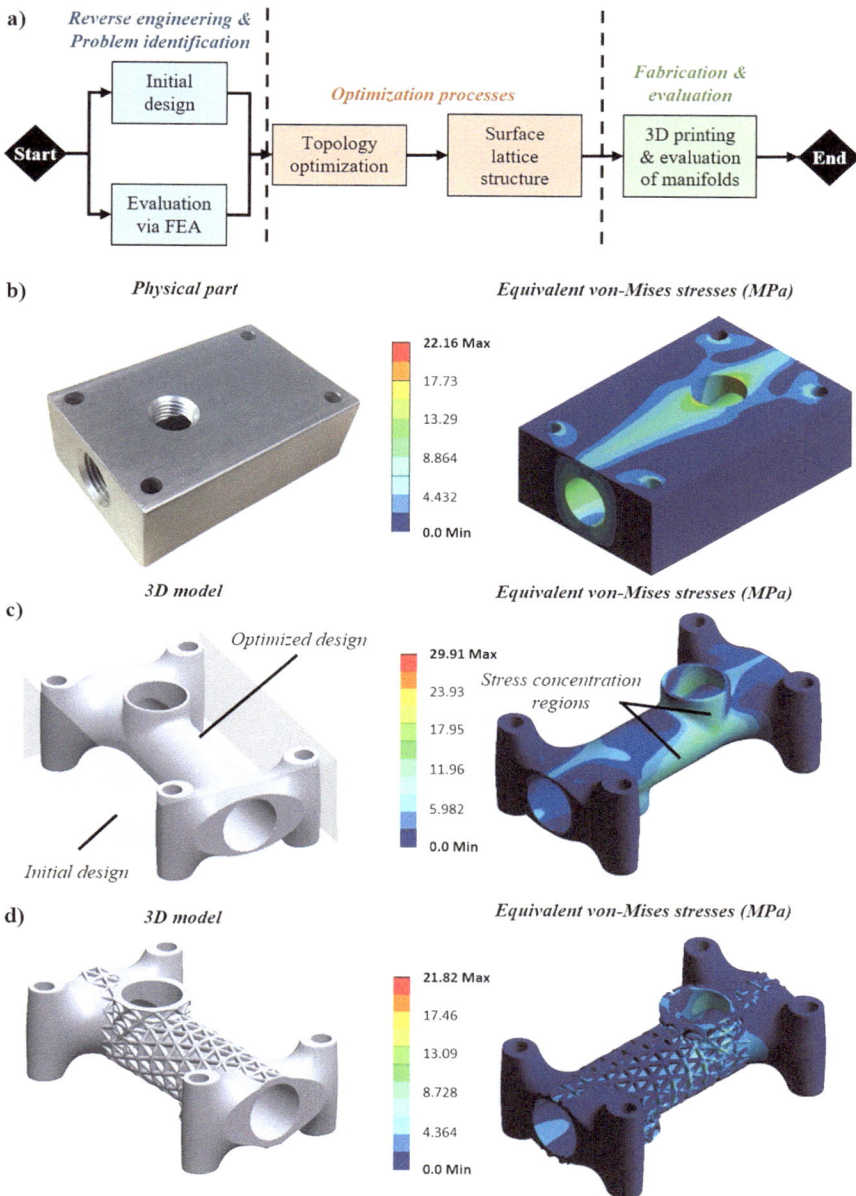

Fig. 6.3 **a** Workflow of the process; 3D models and numerical evaluation for: **b** initial design, **c** element-based approach design, **d** final TO design

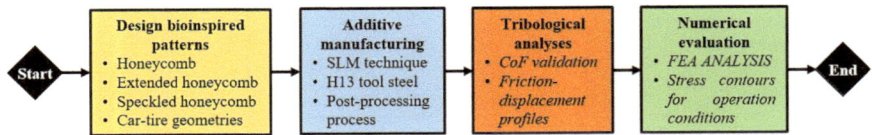

Fig. 6.4 Flowchart of this case study

mechanical response under realistic operation conditions. Finally, the 3D-printed safety gear friction pads were assembled in an existing emergency progressive safety gear system of KLEEMANN Group, providing sufficient functionality. Figure 6.4 presents the flowchart of the current study. It is worth mentioning that in this case study, the topology optimization of the part was performed with the employment of architected materials with the main goal of improving the pad's functionality and brake efficiency.

The developed friction pads, featuring advanced surface geometry, namely the 2.5D architected materials of honeycomb, speckled honeycomb, extended honeycomb, and car-tire-like structure, were produced using the SLM 3D printing technique, as depicted in Fig. 6.5a. The manufacturing procedure was conducted in a low-oxygen environment (<1%) to prevent oxidation. All four designed friction pads were positioned on the 3D printer's platform in a ZYX orientation, as shown in Fig. 6.5b. The entire 3D printing process took approximately 13 h, demonstrating the rapid production capabilities of AM technology. Notably, the 3D-printed parts were structurally flawless, with no visible defects or voids. The 3D-printed pads were weighed with a high-accuracy scale, revealing weights close to the nominal values, indicating minimal porosity in the structures. Following printing, the friction pads underwent post-processing, including support removal and sandblasting, to achieve the final finish. Finally, the friction pads were assembled into the safety gear system, exhibiting proper fit and functionality, as portrayed in Fig. 6.5c.

Based on the tribological results in Table 6.3, the honeycomb and honeycomb-speckled configurations had coefficients of friction (CoF) of 0.317 and 0.372, respectively. These closely matched CoF values suggest a nearly identical pattern in their tribological behavior. However, the speckled variant showed a slightly higher CoF, likely due to the surface speckles. This subtle difference highlights the impact of surface texture on frictional dynamics, emphasizing the need for detailed analysis in tribological studies. Moreover, the extended honeycomb configuration exhibited an average CoF remaining high at around 0.459. In contrast, the car-tire-like configuration showed a significantly higher CoF, reaching the value of 0.549. This notable increase in CoF underscores the unique characteristics of the car-tire-like pattern, where its structural complexities likely contributed to increased frictional interaction. The car-tire-like pattern exhibited the highest CoF among all configurations, establishing it as the configuration with the greatest frictional tendency. These observed behaviors suggest that friction pads with a lower CoF are subjected to higher stress than those with a higher CoF, leading to faster and more abrupt part failure. This conclusion is supported by the finite element analyses (FEAs), as depicted in Fig. 6.6.

Fig. 6.5 **a** Developed 3D-printed friction pads, **b** friction pads inside the building chamber, **c** indicative images of the assembly between the developed friction pads and the elevator safety gear

The commercial friction pad (Fig. 6.6a) and the pad with a honeycomb structure (Fig. 6.6b) demonstrated similar mechanical performance, with equivalent von Mises stress concentrations reaching a maximum of around 170 MPa. However, the friction pad with a speckled honeycomb structure (Fig. 6.6c) experienced the highest stress concentration, reaching 212 MPa, due to localized stress in the honeycomb spots. Specifically, friction pads with extended honeycomb (Fig. 6.6d) and car-tire-like structures (Fig. 6.6e) exhibited lower stress concentrations, effectively distributing the applied loads. Notably, the main stress concentration regions were spotted around the support holes, indicating increased loads on the bolts or screws. The numerical analysis of the friction pads' mechanical behavior demonstrates that topologically optimized pads can withstand friction loads, offering promising results for further experimental testing. In conclusion, configurations with a higher coefficient of friction exhibited better stress distribution, preventing abrupt failures and ensuring a more uniform load distribution across the braking pads, as shown by the FEAs.

Construction reinforcement:

In this case study [8], a systemic approach, that incorporates computer aid and biomimetics, is presented for the topology optimization of 3D-printed clay-based composite mortar reinforced with advanced polymeric bioinspired lattice structures,

Table 6.3 Vertical force evaluation based on CoF and necessary braking force for each structure

Structure	Measured CoF	Braking force	Vertical force
Commercial	0.3	25,506 N	85,020 N
Honeycomb	0.317		80,461 N
Speckled honeycomb	0.372		68,565 N
Extended honeycomb	0.459		55,569 N
Car-tire like	0.549		46,459 N

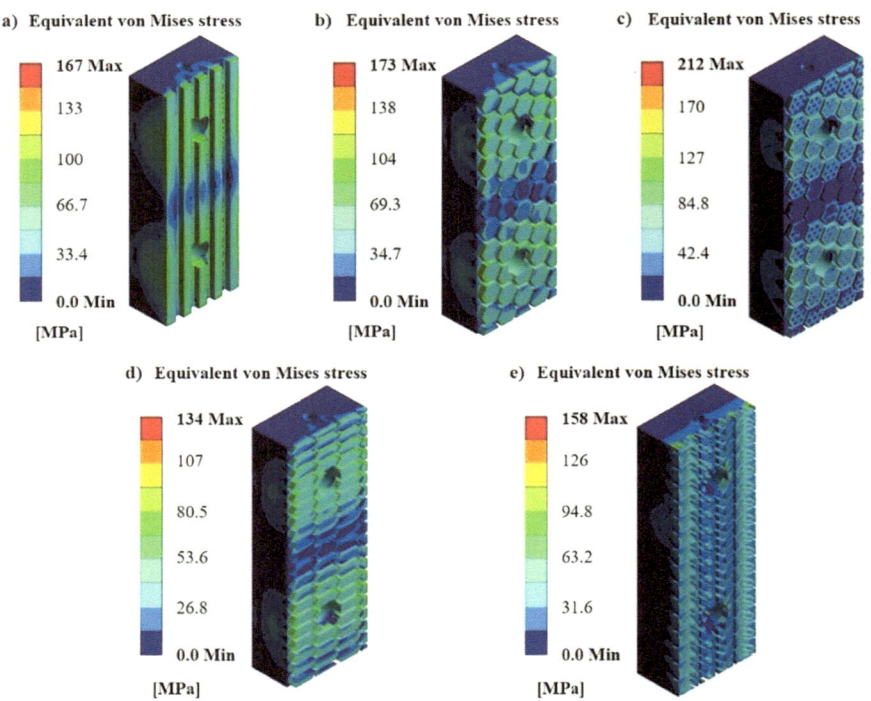

Fig. 6.6 Equivalent von Mises contours for pad structures of: **a** standard **b** honeycomb; **c** speckled honeycomb; **d** extended honeycomb; and **e** car-tire-like

such as honeycombs and Voronoi patterns. These natural lattices were designed and integrated into the 3D-printed clay-based prisms. Then, these configurations were numerically examined as bioinspired lattice applications under three-point bending and realistic loading conditions, by developing appropriate finite element

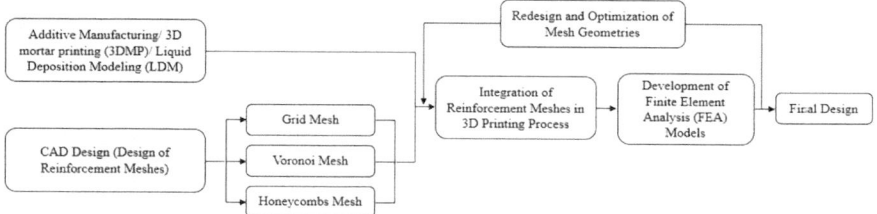

Fig. 6.7 Flowchart of the construction reinforcement case study

models (FEMs). The extracted mechanical responses were observed, and a conceptual redesign of the bioinspired lattice structures was conducted to mitigate high-stress concentration regions and optimize the structures' overall mechanical performance. The topologically optimized bioinspired lattice structures were also examined under the same conditions to verify their mechanical superiority. The results showed that the clay-based prism with honeycomb reinforcement revealed superior mechanical performance compared to the other and is a suitable candidate for further research. The outcomes of this study intend to further research into non-cementitious materials suitable for industrial and civil applications. This study aims to explore non-cementitious materials for industrial and civil applications, focusing on innovative approaches to improve construction materials with low or zero cement content and potentially replace traditional cement-based materials through a topology optimization process. Figure 6.7 shows the workflow of the presented case study.

Regarding the design methodology, a fresh clay-based mixture was produced according to EN1015-11, and then specimens were developed with dimensions of 160 mm × 40 mm × 40 mm each. Based on the aforementioned prisms, bioinspired lattice structures were designed and embedded. These lattice structures are panels consisting of 2.5D architected materials with external dimensions of 160 mm × 40 mm × 2 mm. The first structure was a 2.5D cubic grid, which is the simplest lattice structure yet the most commonly used in civil engineering applications [9]. The second selected architected material was the 2.5D honeycomb-like structure which is a bioinspired structure well-known for its sufficient mechanical performance [7]. For the third architected material, a stochastic structure was developed utilizing the bioinspired Voronoi design algorithm [10]. Through this design methodology, nine distinct designs of clay-based prisms reinforced with bioinspired lattice structures were developed. Figure 6.8 presents indicatively the designed prisms highlighting both their matrix and their reinforcement.

To evaluate the developed clay-based prism reinforced by bioinspired lattice structures, FEMs were developed, examining the compressive and flexural response of each design. Utilizing these prism designs, FEAs were performed for both compression and three-point bending testing. Through these FEAs, the compressive and flexural strengths were evaluated for each specimen coupled with the observed elastic modulus. Table 6.4 presents the basic mechanical properties of the developed design with the relative density of reinforcement at 30%, 50%, and 70%. The acquired data

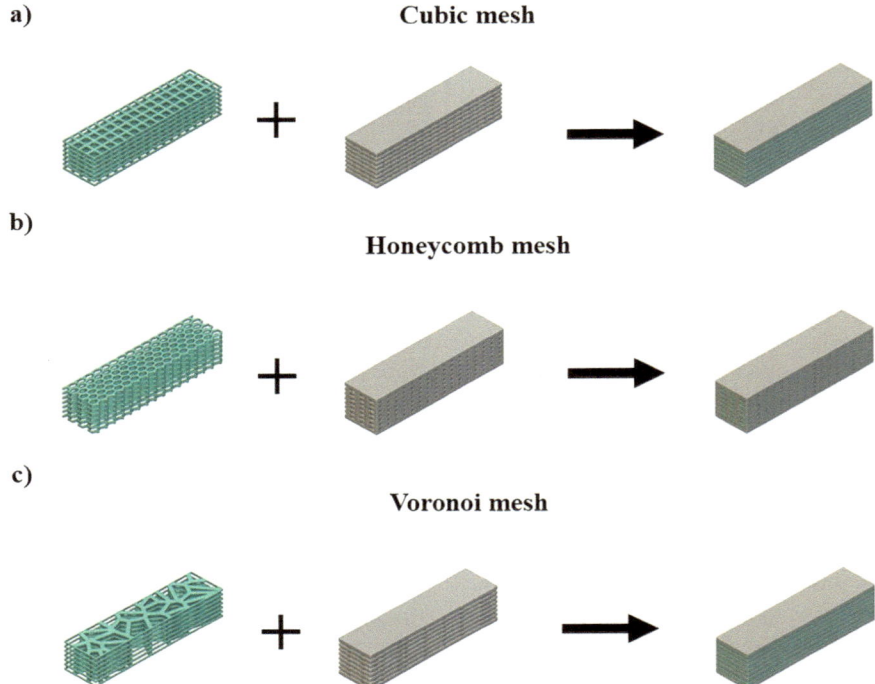

Fig. 6.8 Developed prism designs for: **a** cubic mesh, **b** honeycomb mesh, and **c** Voronoi mesh

showed that the clay-based prisms with the honeycomb reinforcement revealed the highest stiffness and strength, followed by the ones with cubic reinforcement and the Voronoi structure.

Similar findings were revealed in the bending evaluation where reinforced clay-based prisms exhibited higher flexural strength, i.e. maximum force, than the pure clay-based prisms, as it is listed in Table 6.5. Basic mechanical properties from bending testing. Again, honeycomb reinforcement revealed the highest flexural strength at 3.41 MPa, reaching a maximum bending strength of 1553 N. On the other hand, cubic and Voronoi reinforcement have similar strengths with cubic structure showing a little higher flexural strength in general.

Finally, in Fig. 6.9, indicative images of FEA results from bending evaluation are depicted focusing separately on the two elements of the developed prism, namely the matrix and the reinforcement.

Table 6.4 Basic mechanical properties from compression testing

Specimens	Elastic modulus	Yield strength	Peak strength
No reinforcement			
Pure clay-based (MPa)	4710	1.12	1.57
Cubic mesh			
Clay-based with 30% PLA reinforcement (MPa)	4100	1.26	1.69
Clay-based with 50% PLA reinforcement (MPa)	3648	1.55	2.01
Clay-based with 70% PLA reinforcement (MPa)	3010	1.76	2.34
Honeycomb mesh			
Clay-based with 30% PLA reinforcement (MPa)	4020	1.58	1.94
Clay-based with 50% PLA reinforcement (MPa)	3880	1.72	2.32
Clay-based with 70% PLA reinforcement (MPa)	3650	1.83	2.69
Voronoi mesh			
Clay-based with 30% PLA reinforcement (MPa)	4130	1.05	1.53
Clay-based with 50% PLA reinforcement (MPa)	3530	1.32	1.67
Clay-based with 70% PLA reinforcement (MPa)	3140	1.56	1.95

Table 6.5 Basic mechanical properties from bending testing

Specimens	Maximum force	Flexural strength
No reinforcement		
Pure clay-based	431 N	1.01 MPa
Cubic mesh		
Clay-based with 30% PLA reinforcement	517 N	1.21 MPa
Clay-based with 50% PLA reinforcement	746 N	1.75 MPa
Clay-based with 70% PLA reinforcement	1010 N	2.37 MPa
Honeycomb mesh		
Clay-based with 30% PLA reinforcement	751 N	1.76 MPa
Clay-based with 50% PLA reinforcement	1094 N	2.56 MPa
Clay-based with 70% PLA reinforcement	1553 N	3.41 MPa
Voronoi mesh		
Clay-based with 30% PLA reinforcement	440 N	1.03 MPa
Clay-based with 50% PLA reinforcement	654 N	1.53 MPa
Clay-based with 70% PLA reinforcement	937 N	2.19 MPa

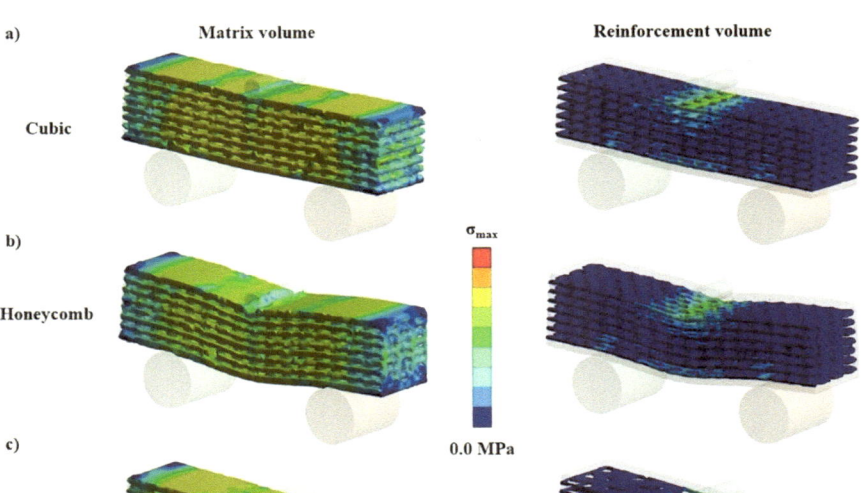

Fig. 6.9 Indicative equivalent von Mises stress contours of reinforced specimens for: **a** cubic, **b** honeycomb, and **c** Voronoi prisms

6.2.2 3D Architected Materials Applications

The utilization of 3D architected materials is way more complex in terms of both design and manufacturability. Therefore, their usage is limited in applications where their impact is sufficiently high, such as biomechanical and impact (crashworthiness) applications. In this context, the current book aims to present indicative case studies from both conducted studies by the authors and also from the existing literature.

Tibial Scaffold and Implant:

The first case study on topology optimization utilizing 3D architected materials focuses on the development of novel tibial scaffolds. In detail, this study developed and fabricated novel customized tibial scaffolds utilizing advanced design methods, bioactive polymers, and AM techniques [11]. The developed scaffolds are suitable for tibial shaft injuries with extensive bone defects due to their "bone-brick" configuration. The novelty of the topology optimization for the tibial scaffold lies in the implementation of advanced internal scaffold designs, which enhance biocompatibility and bio-functionality. These designs incorporate high porosity and imitate the bone's diffusion canals [12]. Figure 6.10 illustrates the flowchart of the topology optimization procedure. The study examined several strategies for biomimetic diffusion canals and found that the optimal design for Volkmann-like canals in terms of mechanical performance and related surface area is an elliptical cross-section, while circular cross-sections are retained for Haversian-like canals. The circular cross-section canals had a diameter of 1 mm and followed the perimeter of the scaffold with 16 total vertical canals, while the elliptical cross-section canals had a size of 1 mm for the minor axis and 2 mm for the major axis with a total of 5 horizontal canals [12]. In addition, the Schwarz Diamond structure and SD&FCC hybrid cellular material were chosen as the best architected materials for this application. The relative density of the lattice structures was chosen at 20% which is similar to the bone porosity (80%) [11]. Furthermore, the length of the unit cells was set at 10 mm and the wall/strut thickness was set at 0.858 mm and 0.745 mm for Schwarz Diamond and SD&FCC structure, respectively. Finally, as construction materials the biodegradable and bioabsorbable materials PLA and PCL were selected. The final designs were examined both experimentally and numerically to verify their structural integrity under the load of activities of daily living (ADL).

In Fig. 6.11(a–c), the final designs of the developed scaffolds are depicted by following the aforementioned flowchart. To evaluate the accuracy and functionality of the designed scaffolds, a surgical tibia model made from bone-like material was utilized from the SYNBONE™, which was fabricated in a perfect replica of the model from CT scans. The developed socket configurations of the scaffolds were found to fit perfectly within the bone fragments, aligning and connecting the proper positions for tissue healing. It is important to note that the FFF 3D printer, used for manufacturing these complex geometry of diffusion canals and lattice structures for both materials (Fig. 6.11b), had sufficient accuracy. In addition, Fig. 6.11c shows indicative images of the internal structure of scaffolds made with PLA and PCL captured by the digital microscope system Leica DMS 1000. The PLA scaffolds exhibited adequate dimensional accuracy without any visible structural flaws resulting from the 3D printing process. Conversely, the PCL scaffolds exhibited poor dimensional accuracy, especially in the diffusion canals. Additionally, 3D printing defects, such as material discontinuities, were observed along the body of the lattice structure. These issues were mainly attributed to the low printability and high elasticity of the PCL raw material. It is noteworthy that the scaffold could be fixed with bone with four screws, which were bioabsorbable and suitable for medical purposes.

Table 6.6 presents the mechanical properties values of the designed scaffolds, including the maximum force, force at yield, stiffness, and elongation at yield, as they were derived through compressive mechanical experiments. The maximum operation force that a tibia scaffold should withstand to be functional is 5300 N for a human with a body weight of 75 kg (for intense exercise) [13]. The PLA scaffolds exhibited remarkable strength with respect to the maximum operation force, achieving twice the load of the scaffold with SD structure and more than triple that of the scaffold with SD&FCC structure. On the other hand, none of the developed PCL scaffolds managed to overcome the operational force. However, the scaffold with SD&FCC structure withstood a maximum force of 4800 N, leading to the conclusion that minor

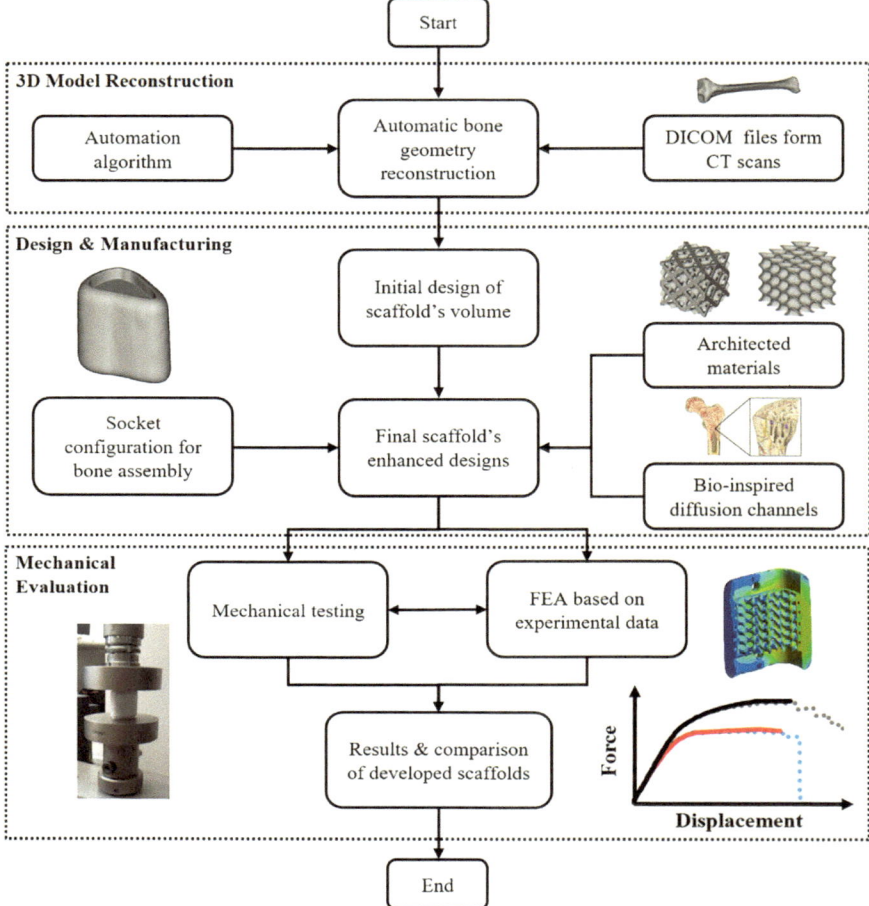

Fig. 6.10 Flowchart of the topology optimization procedure for the bone-brick scaffold

adjustments to the raw material, such as the introduction of nanocomposites or modifications to the scaffold design, such as an increase in the diffusion canal region, could result in a sufficient mechanical response. As expected, PLA scaffolds demonstrated over six times higher stiffness compared to the PCL scaffolds. However, the increased elasticity of the PCL scaffold could be a comprehensive advantage in biomechanical applications due to its ability to absorb a high amount of energy, by enabling larger displacements in the case of extreme loading conditions, such as impact or crashes.

The second biomechanical case study presented in this book concerns the development of a novel tibial implant consisting of advanced architected materials. The tibial implant is a critical component used in total knee replacement surgery. The selection of the proper architected materials occurred based on the existing architected

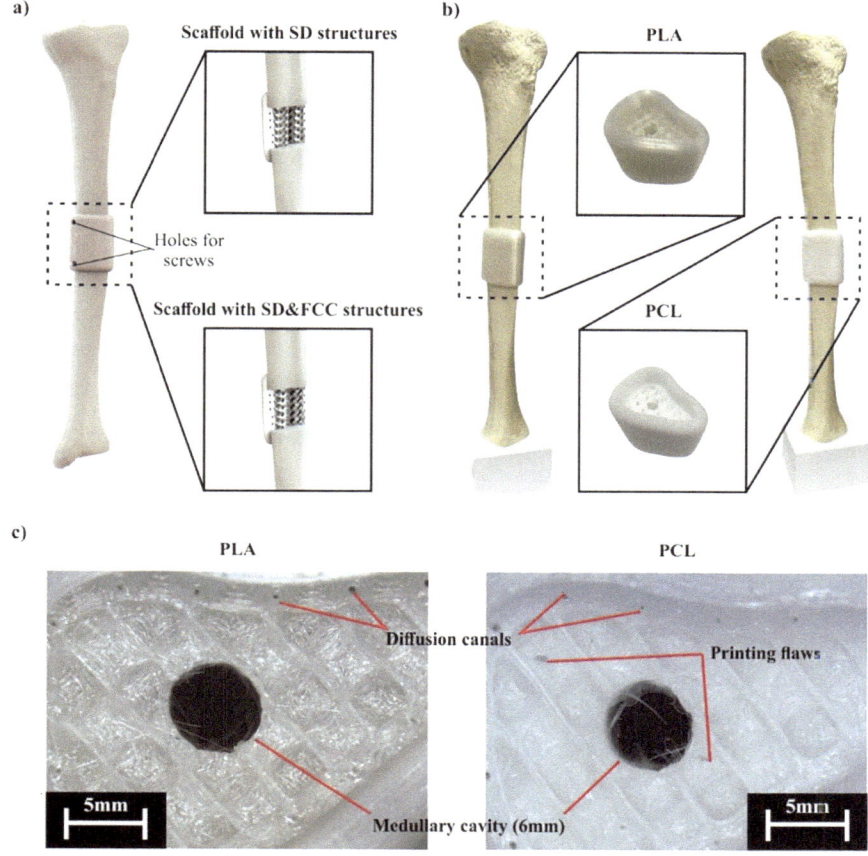

Fig. 6.11 Depiction of the scaffold assembly coupled with superimposed images from the internal structure: **a** 3D render digital model, **b** physical 3D-printed scaffolds, and **c** indicative images of scaffolds internal structure for PLA and PCL

Table 6.6 Values of the mechanical performance for the developed scaffolds

Properties	Scaffold materials			
	PLA		PCL	
	Lattice structures			
	SD	SD&FCC	SD	SD&FCC
Max. force (N)	12,650	17,704	3892	4803
Force at yield (N)	9343	12,338	2792	3528
Stiffness (N/mm)	5840	5890	845	930
Elongation at yield (mm)	1.6	2.1	3.3	3.8

materials and the developed hybrid cellular structures. Since these implants are typically made from metals using the SLM technique, the material selection process was performed as a result of the evaluation of each structure's mechanical performance through the methodology established for the SLS RVEs in previous chapters. Once the suitable architected materials were identified, the next step was to define the lattice structure application. Having the final design of the implant structure completed, the 3D printing process occurred via SLM AM technology. Finally, both the 3D model of the implant and the physical 3D-printed part were evaluated with experimental analysis. Figure 6.12 presents followed workflow for this case study. To summarize, the aim of this case study was to develop an innovative lightweight structure for a tibial implant for total knee replacement application. The novelty of this topology optimization process offers sufficient structural integrity with maximum porosity with the employment of advanced hybrid architected material. Porosity possesses a crucial role in the diffusion of blood and nutrients leading to higher biocompatibility of the implant. Moreover, the high porosity of the applied architected material led also to a low relative density, resulting in an overall structure with strength and stiffness closer to that of the bone. This could significantly reduce the stress-shielding effect that is commonly observed in metallic bone implants. Finally, the employment of conformal and functionally graded architected materials in a specific region can increase the friction between the bone and the implant reducing the wear effect. Also, the functional gradation secures the necessary structural integrity of the implant. Therefore, it is obvious that this topology optimization process offers multiple advantages on the final design achieving superior performance compared to the topology optimization process applying element-based approach.

In detail, SD&FCC architected materials were utilized for the replacement of the solid tibial plate. In order to achieve this, this specific region of the implant was isolated and redesigned using the desired architected material with a unit cell length of 5.2 mm (plate's thickness) and a wall/strut thickness of 0.52 mm, as it is illustrated in Fig. 6.13a. On the other hand, for the Octet structure, a conformalization of the structure was employed, as it was shown in Fig. 6.13b. Furthermore, the Octet

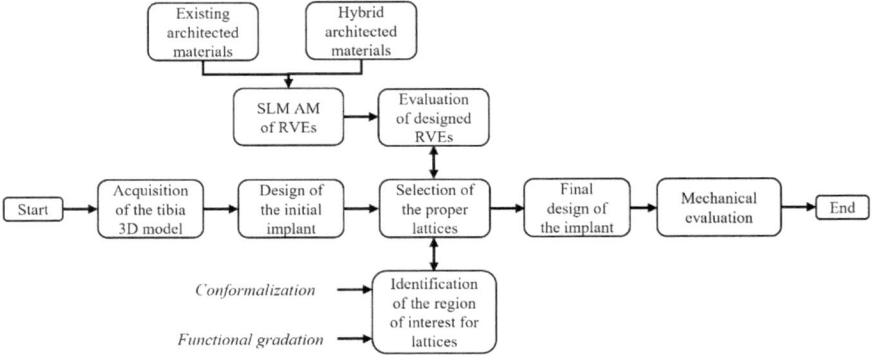

Fig. 6.12 Flowchart of the tibial implant case study

structure had a functionally graded relative density with values that were increased from the external surface to the center resulting in a complete bulk design after one point. The external layer of the unit cells was constructed with a minimum possible strut thickness of 200 µm to achieve high elasticity for the final structure. In addition, the developed tibial implant's final design is presented in Fig. 6.13c. Figure 6.13d shows the assembled configuration between the developed tibial implant and a surgical tibia model. It is apparent that the designed implant fits perfectly with the geometry of the physical tibia bone. Finally, it is worth noting that a superimposed image of the Octet microstructure is also depicted, with an optical stereoscope, as portrayed in Fig. 6.13e.

The developed tibial implant was also evaluated with compressive experiments in terms of final mass reduction and overall strength. Table 6.7 lists the most important values of these properties. According to this table, the topologically optimized implant had a sufficiently lower mass than the initial design of 73.2 g, resulting in an overall mass reduction of 58.8%. Moreover, the maximum vertical compressive force, that the developed implant can withstand, was observed at 15792N way beyond the loads from APLs, but relatively close to the strength of a tibia bone from an adult male [14]. To conclude, the TO tibial implant revealed strength and stiffness close to the bone, thereby minimizing the potential stress-shielding effect. Furthermore, the main porosity of the implant was at 70%, except for specific pre-selected solid regions. This high porosity enhances the biocompatibility of the implant and facilitates the osseointegration process. Finally, the applied conformal Octet structure on the stem increases the friction force between the implant and the bone leading to reduced relative movement and minimizing the wear effect.

Sports Helmet:

The final topology optimization case study in this book involves the development of a lightweight sports helmet. This helmet employed functionally graded 3D architected materials in a sandwich-like configuration to achieve high strength and improve energy absorption capabilities. Figure 6.14 depicts the methodology applied in the development of the sports helmet.

The objective of this mechanical case study was to develop a lightweight sports helmet for various athletic activities (climbing, riding, etc.) exploiting the increased energy absorption of advanced architected materials, such as hybrid and functionally graded, in order to significantly reduce the probability of injury. The novelty of the current topology optimization process offers a sports helmet design with enhanced structural integrity and increased energy absorption capability utilizing only one construction material. Hence, a significant reduction in the probability of injury from impact was revealed compared to a solid helmet.

The developed helmet was designed to exploit the dimensions and morphology of the dummy head based on the International Society for the Advancement of Kinanthropometry (ISAK) and international anthropometric standards [15] in order to be compatible with the head of an average adult male. Besides the specific geometry, a sports helmet should fulfill the following conditions. First, it must sufficiently cover the head area above the ears and from the upper part of the neck up to the

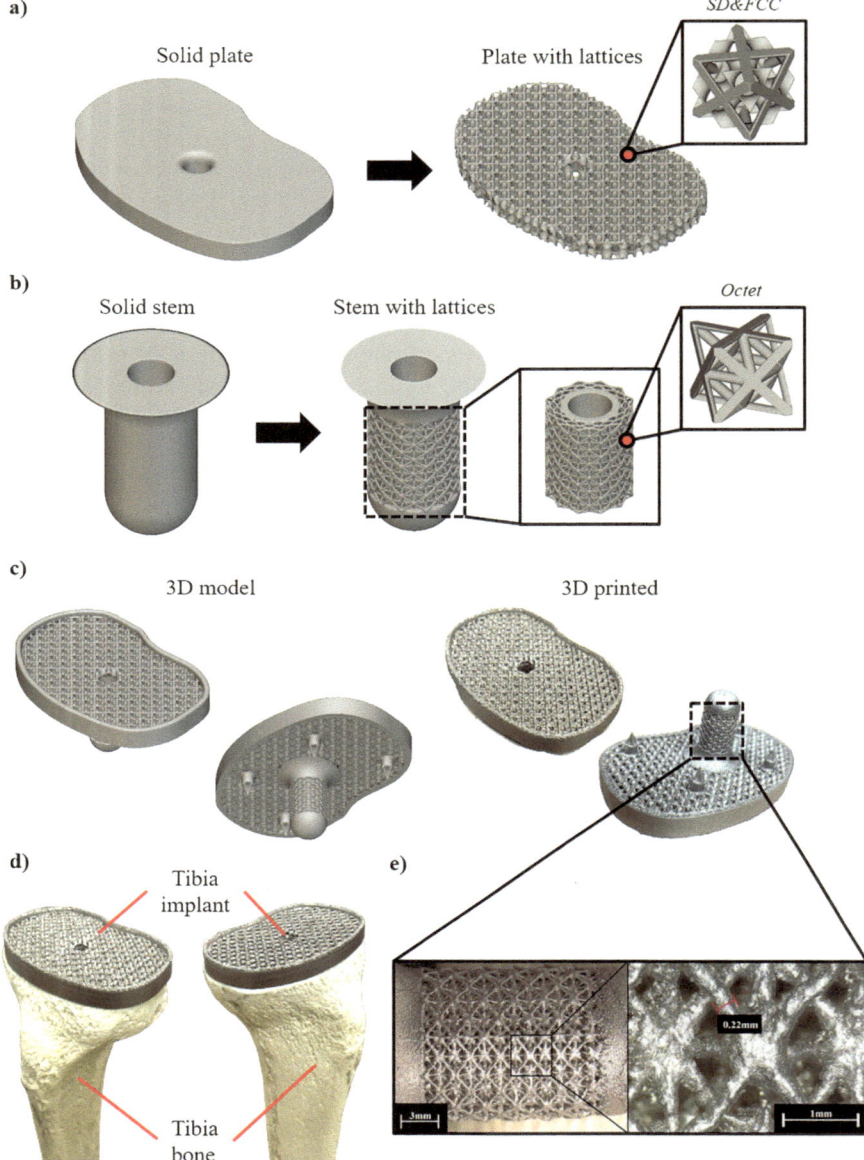

Fig. 6.13 **a** Redesign of the tibial plate, **b** redesign of the tibial stem, **c** images of the developed 3D model and the 3D-printed tibial implant, **d** assembly configuration of the tibial implant, and **e** superimposed images of the conformal Octet lattice structure

Table 6.7 Values for mass and strength for TO tibial implant

Design	Initial mass	Final mass	Mass reduction	Maxstrength
Tibial implant	177.7 g	73.2 g	58.8%	15792N

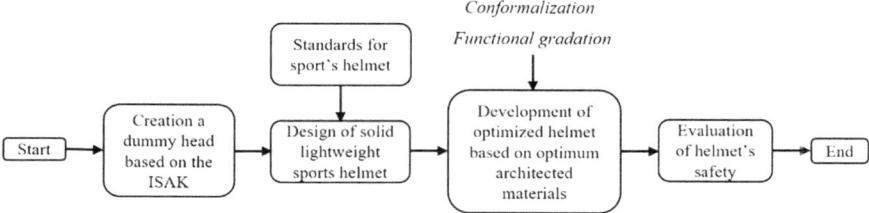

Fig. 6.14 Flowchart of the sports helmet application

forehead including the temples, top, sides, and back of the head. Furthermore, a sports helmet should be lightweight, i.e. mass below 2 kg, in order to offer comfort, increased performance, maneuverability, and safety. Moreover, it is essential for a sports helmet to provide the necessary ventilation for the head to reduce the risk of head overheating during the activity. Finally, helmets should be constructed of rigid and durable materials with enhanced energy absorption capabilities. Taking into account these, in the context of this case study, a sports helmet was designed to fulfill all the above-mentioned specifications, with external dimensions of 276 mm × 233 mm × 138 mm. This helmet was designed in order to be manufactured with a rigid polymer with enhanced energy absorption performance, namely PA12. The above-presented helmet had a bulk design with numerous ventilation holes, however, the overall weight of the object was 1.67 kg, which is marginally below the specifications. Also, the energy absorbability of the developed helmet is questionable due to the bulkiness of the structure. Thus, it was decided to replace the solid regions between the inner and the outer surfaces with advanced architected materials employing a topology optimization process. In detail, functionally graded and conformal lattice structures were designed to match the helmet's shape and achieve optimal performance. Two different lattice structures were derived employing two different architected materials, i.e. the SD and SD&FCC. These architected materials were selected for their strength, energy absorption performance, and the ability to be transformed into IPCs, which further improve their energy absorption capability. The mean relative density of the functionally graded and conformal lattice structures was chosen at 40%, with linear gradation. More specifically, the denser structures were close to the head (helmet's inner surface) by up to 60% and the lighter ones were positioned close to the helmet's outer surface with a relative density of almost 20%. As it was presented in the previous chapter, this configuration revealed the optimum performance in terms of energy absorption. The employed architected materials had a constant mean unit cell length of 10 mm and the gradation was achieved through the wall/strut thickness. The wall thickness for the SD architected materials ranged from

0.86 mm to 1.58 mm and the wall/strut thickness for the SD&FCC hybrid architected material ranged from 0.7/0.8 mm (SD/FCC) to 2.5/1.5 mm (SD/FCC). Figure 6.15(a and b) shows the developed designs that were conducted in the nTopology™ design platform. In these images, it is obvious that the employed architected materials had a sandwich-like configuration with inner and outer surfaces of 3 mm. Moreover, the superimposed images present the conformal and functionally graded nature of the lattice structures. In addition, the lattice structures are uncovered in the sides in order to easily release the applied stress and allow the filling of the structure with expanding foam (polyurethane foam) for further structure improvement. The overall volume and dimension of the designed helmet remained unchanged. However, a significant weight reduction was achieved with an overall weight of 0.86 kg for both designs due to the same mean relative density for both lattice structures.

The evaluation of the developed helmet designs was performed through numerical studies utilizing the explicit dynamic module of the ANSYS™ platform. In detail, the analyses simulated the vertical impacts of the helmets in a solid wall with a linear velocity of 11 m/s (worst-case scenario). The impacts were simulated until a stable rebound velocity was achieved, with an overall T of 40 ms.

The results of the numerical analyses are illustrated in Fig. 6.16. In detail, Fig. 6.16 presents the stress contour for the developed helmet design, i.e. solid, SD, and SD&FCC, at three different impact stages, the initial where there are zero displacements (left side), the impact where the maximum acceleration ($v = 0$ m/s) is observed (middle) and the rebound stage where the rebound velocity has been stabilized (right side). The following qualitative conclusions can be derived concerning all the conducted impact analyses. Initially, before the impact occurred all three analyses had the same metrics. During the second stage, the maximum stresses are observed coupled with the maximum strains of the structure. Moreover, as was expected the regions of the impact experienced the maximum stresses that were distributed within the structure's volume. Finally, in the rebound stage, internal stresses from the impact were released in the rest of the structure with significantly lower magnitude, furthermore, a constant rebound velocity occurred in the opposite direction of the initial velocity. This rebound velocity of each structure is directly linked with their absorbed energy which is dependent on the structure's stresses and strains.

More specifically, the solid helmet showed better performance in terms of structural integrity, as it was revealed the lower stress concentration inside its structure with a maximum value of 24.62 MPa. However, this led to increased elastic strain released during the rebound stage with a rebound velocity of 10.9 m/s resulting in decreased mechanical energy absorption. On the other hand, the helmet with the SD structure revealed maximum stress of 33.86 MPa with increased plastic strains for the internal lattice structure and the fraction of the external structure as it surpassed the material's ultimate strength (32 MPa). These resulted in compromised structural integrity and a reduced rebound velocity of 10.2 ms/. A similar pattern with the helmet with the SD structure was observed for the helmet with the SD&FCC structure. This helmet's design revealed a maximum stress of 30.44 MPa and a rebound velocity of 9.6 m/s. The unique characteristic of this case is that the hybrid architected material experienced significant fracture on the low relative density region for the FCC part

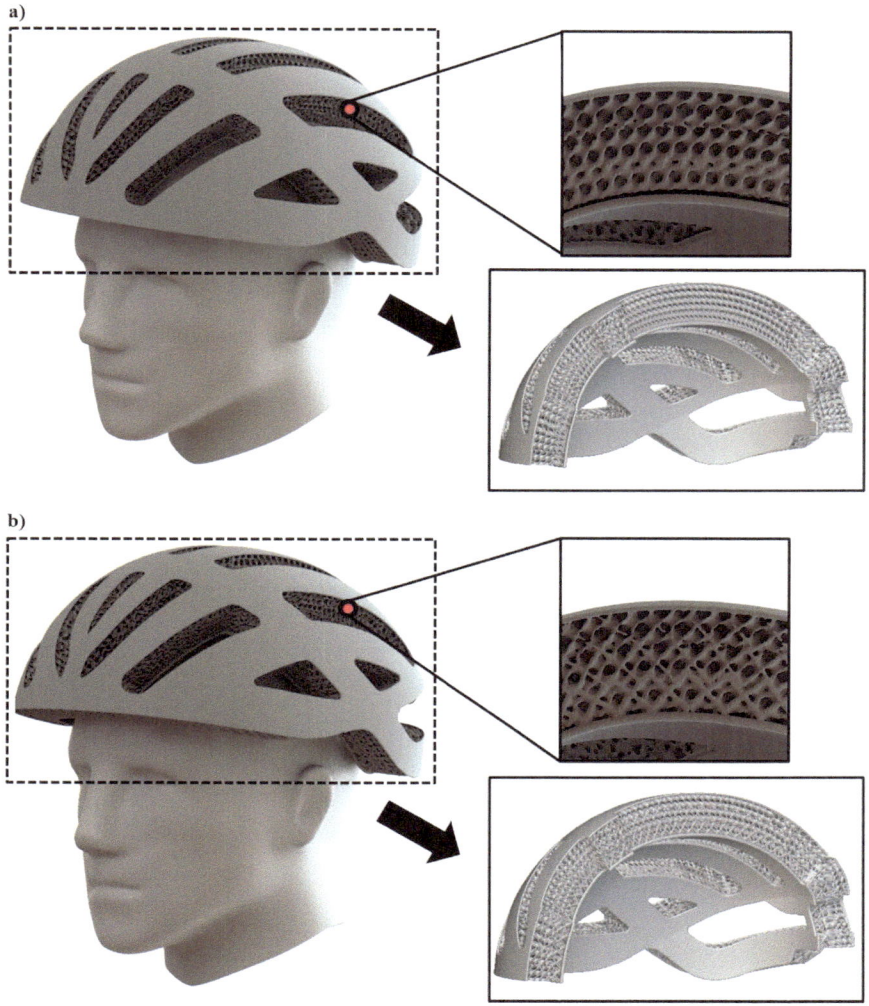

Fig. 6.15 Design helmet with conformal and functionally graded: **a** SD and **b** SD&FCC architected materials

of the structure with complete breakage of the strut which led to extensive energy absorption. In contrast, the SD part of the hybrid structure maintained the structural integrity of the overall helmet.

Having described qualitative impact characteristics and the stress distribution for the examined designs of this case study, the next step is the evaluation of each design through specific performance indicators, namely the average acceleration, the head injury criterion (HIC) values, the factor of safety (FOS) values, and the weight of each structure [16–18]. Table 6.8 lists the values of each of these indications for

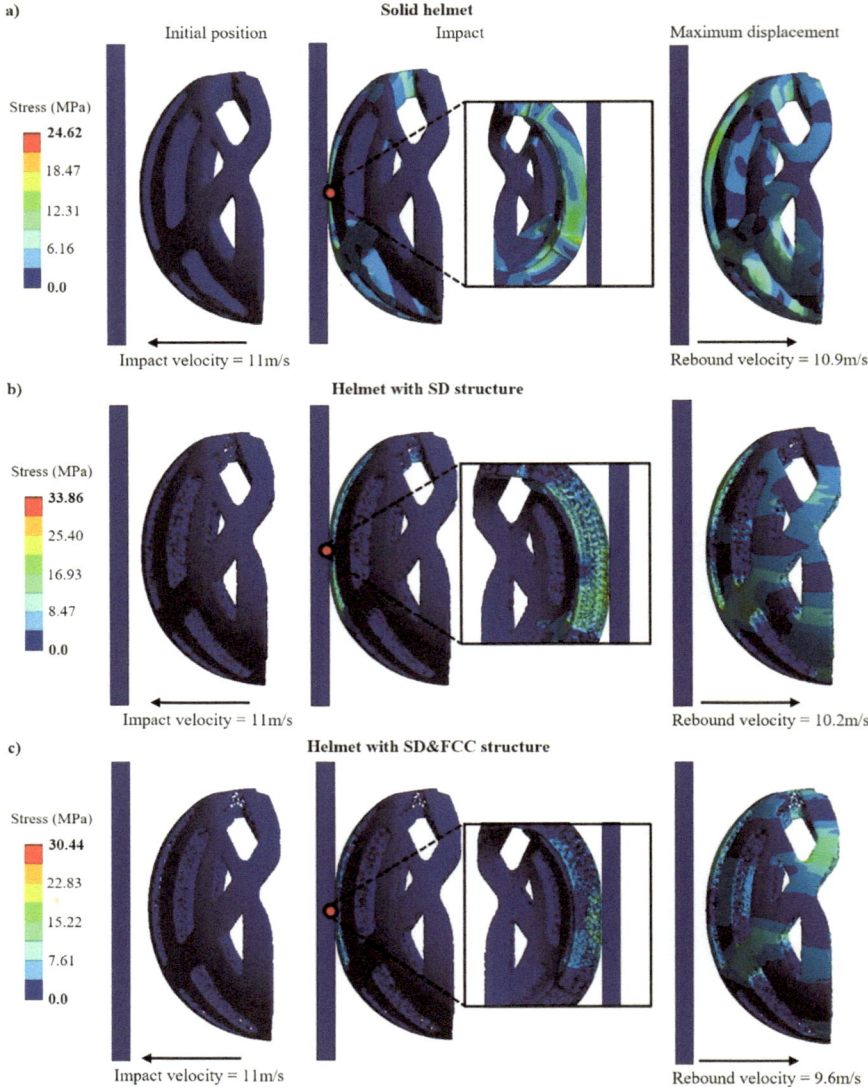

Fig. 6.16 Stress contours for the developed helmet designs of: **a** solid, **b** SD, and **c** SD&FCC

all examined helmet designs. First, the designs with the integrated lattice structure led to a mass reduction of 48.5% achieving final lightweight designs. Then, through the FEA, the FOS values for each design were derived, i.e. how much stronger the structure is, compared to what it needs to be for the applied load. All three designs exceeded their yield point due to the increased initial velocity, but only the helmet with the SD structure surpassed the ultimate strength of the construction material.

The other two values, i.e. the average acceleration and HIC, concerning the overall performance of each helmet and their ability to protect the human's head in case of impact, were numerically evaluated. Both values are commonly known to be as low as possible. According to Table 6.8, the design with the lattice structure revealed lower average accelerations and HIC values due to their increased energy absorption. One last result of this case study is the representation of the calculated HIC value on the AIS diagram (Fig. 6.17). Based on Fig. 6.17, the helmet with SD&FCC almost eliminated the probability of fatal injury and significantly increased the probability of moderate injury for a vertical crash with an impact velocity of 11 m/s ($\approx$ 40 km). Finally, the helmet with the integrated lattice structures can be further improved by employing IPCs utilizing the developed lattice structures as core phases. This could result in a HIC value of 1000 which is a safety landmark for the protective equipment.

Other Applications:

Besides the conducted research on case studies of topology optimization with architected materials from the authors, in this section, a series of case studies derived from the published literature is presented. The initial utilization of architected materials and lattice structures was for the purpose of reducing weight in the automotive and aeronautics industries, due to their high porosity [19–21]. In detail, these materials were initially employed as cores in sandwich-like structures, to achieve a reduction in mass while maintaining strength and stiffness [22, 23], as it is presented in Fig. 6.18a. With the advent of AM technologies, the employment of architected

Table 6.8 Values of the basic performance indicators for the developed designs

Design	Average acceleration (gs)	HIC	FOS	Weight (kg)
Solid	89.33	1885	0.93	1.67
SD	86.47	1738	0.68	0.86
SD&FCC	84.02	1618	0.76	0.86

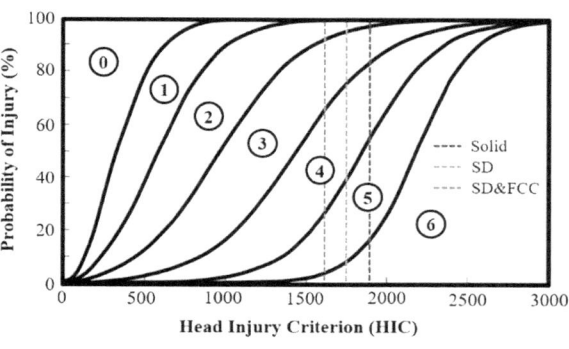

Abbreviated Injury Scale (AIS)

Region 0: No injury
Region 1: Minor injury
Region 2: Moderate injury
Region 3: Serious injury
Region 4: Severe injury
Region 5: Critical injury
Region 6: Fatal injury

Fig. 6.17 HIC values of the developed design compared with the AIS

materials has shifted toward the topology optimization of existing structural components. For these applications, architected materials displaying stretching-dominated
behavior are typically chosen for their enhanced stiffness and peak strength. As a
result, the solid portions of the existing components are replaced with lattice structures that possess a specific geometry and relative density, in order to achieve the
desired mechanical properties.

In addition to their structural optimization capabilities, architected materials also
exhibit enhanced energy absorption performance, as previously discussed. Their
ability to absorb significant amounts of mechanical energy presents a variety of
opportunities for applications, such as crashworthiness and impact resistance. For
example, lattice structures are used for packaging, as they can absorb energy and
loads to protect the product [24]. Similarly, lattice structures have been incorporated
into the protective equipment of athletes, such as helmets and splints. Furthermore,
due to their high crashworthiness, architected materials have been proposed as a
protective shield in aerospace vehicles [25]. Furthermore, architected materials made
of elastomeric materials can be utilized as energy absorbers [26].

Besides their mechanical behavior, architected materials possess two key physical
characteristics that provide them with a competitive advantage for a wide range of
applications. These characteristics are high porosity and high surface area to volume
ratio, particularly in sheet-architected materials. The high porosity of these materials
results in a partially filled overall volume, with the remaining volume filled with air
or other materials. This leads to an increase in thermal insulation, as the existence
of air between the pores of the structure operates as an insulator and reduces the
thermal conductivity. This feature has led to the employment of architected materials,
particularly honeycombs, in the aerospace industry for thermal insulation in booster
rockets for space shuttles. In addition, the presence of air in the closed-cell architected
materials imparts the property of buoyancy, which, in combination with high energy
absorption and adequate mechanical strength, makes them suitable materials for
the construction of lightweight boats and canoes. Furthermore, due to the same
principle, highly porous architected materials can be utilized in acoustic insulation
applications, as the cellular structure in conjunction with air pores results in a high
damping capacity of the structure, functioning as a sound absorber, [27, 28] as it
is illustrated in Fig. 6.18b. On the other hand, given that a significant proportion
of volume is often unoccupied or filled with air, researchers investigate the use of
3D-printed biomaterials utilizing 3D bio-printers or other AM methods, such as FFF
and SLS, to create bio-polymers.

The previously presented case studies demonstrate that incorporating pores into
a structure can accelerate tissue regeneration. The presence of large pores and chan-
nels facilitates the diffusion of blood and nutrients, creating an ideal environment for
growth factors that promote cell regeneration. Moreover, the high porosity and high
surface area to volume ratio of these architected materials provide an ideal environment for the placement of growth factors promoting cell regeneration. Additionally,
the high surface area to volume ratio of architected materials, particularly in sheet
lattices, is a distinctive feature that can accelerate chemical and physical processes,
due to the fact that are contingent upon contact surface area or contact length, such

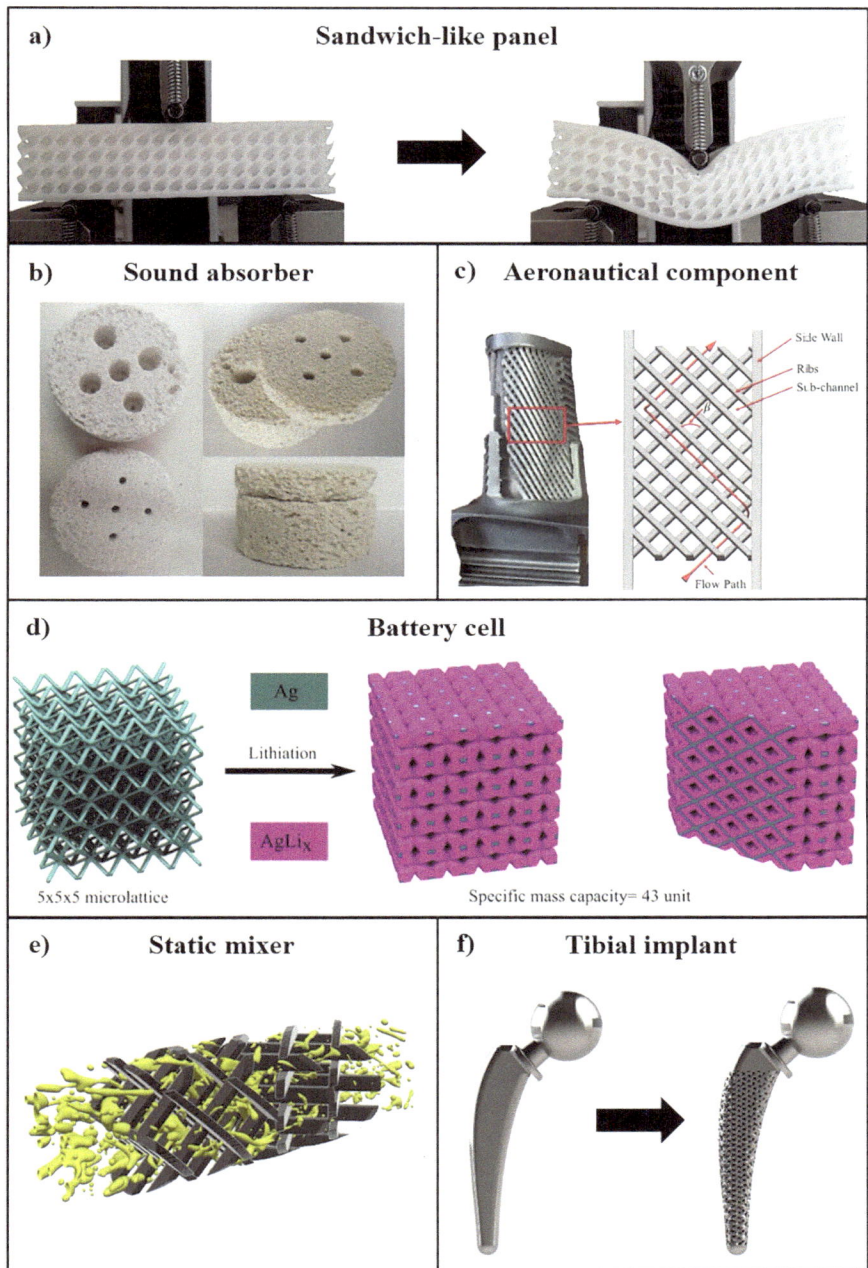

Fig. 6.18 Application of TO with architected materials for: **a** sandwich-like panel [22], **b** sound absorber [28], **c** aeronautical component [29] (Copyrights 2019 Elsevier), **d** battery cell [31] (Copyrights 2018 Elsevier), **e** static mixer [32] (Copyrights 2020 Elsevier), and **f** hip implant [34]

as heat transfer and catalysis, such as the turbine blade configuration that is shown in Fig. 6.18c. More specifically, the larger the surface area or the length of the surface area the faster the physical or chemical process occurs. Therefore, architected materials with a large surface area to volume ratio, such as sheet-TPMS, are employed for heat exchangers and temperature control applications that are fabricated using the SLM additive manufacturing method [29]. Additionally, this principle applies to chemical reactions, making architected materials suitable for catalysis and battery applications [30, 31] as it is depicted in Fig. 6.18d.

Other potential applications of architected materials include filters and static mixers. These applications are derived from a combination of characteristics, such as high porosity, high surface area to volume ratio, and high geometrical complexity. Ceramic architected materials have been used as filters in metal casting due to their ability to trap inclusions with the structure. Static mixers, which are mixers without any moving parts, mainly consist of sheet-architected materials and mix two or more fluids through their geometry [32] (Fig. 6.18e). Additionally, architected materials exhibit high friction forces due to their unique shape and multiple surfaces. As a result, they have been utilized as carriers of fluids with high viscosity, which can control or stop the flow of fluid without the need for external force, or as components with high friction for the secure assembly of components. One particularly successful example of this is the high-performance Trilock hip implant configuration fabricated using the SLS additive manufacturing technique [33]. Figure 6.18e illustrates indicative images of this implant configuration.

References

1. N. Kladovasilakis, T.Bountourelis, K. Tsongas, D. Tzetzis, Computational Investigation of a Tibial Implant Using Topology Optimization and Finite Element Analysis. *Technologies* (2023), 11, 58. https://doi.org/10.3390/technologies11020058
2. N. Kladovasilakis, G. Kosmidis, P. Kyratsis, D. Tzetzis, *Topology Optimization Utilizing Density-Based Approach for Additive Manufactured Components: A Case Study of an Automotive Brake Caliper* (2023) pp. 91–106. https://doi.org/10.1007/978-3-031-21167-6_4
3. T.E. Bountourelis, N. Kladovasilakis, K. Tsongas, P. Kyratsis, D. Tzetzis, Finite element analysis and topology optimization of a high-performance lower prosthetic limb. Int. J. Modern Manuf. Technol. **XV**(1) (2023). https://doi.org/10.54684/ijmmt.2023.15.1.148
4. https://grabcad.com/challenges/ge-jet-engine-bracket-challenge. Accessed 20 August 2024
5. A. W. Gebisa, H. G. Lemu, A Case Study on Topology Optimized Design for Additive Manufacturing. IOP Conf Ser Mater Sci Eng (2017), 276, 012026. https://doi.org/10.1088/1757-899X/276/1/012026
6. N. Kladovasilakis, Design for additive manufacturing of topologically optimized air manifold. Int. J. Modern Manuf. Technol. **XVI**(3) (2024). https://doi.org/10.54684/ijmmt.2024.16.3.136
7. N. Kladovasilakis, E.M. Pechlivani, I.K. Sfampa, K. Tsongas, A. Korlos, C. David, D. Tzovaras, Metal 3D-printed bioinspired lattice elevator braking pads for enhanced dynamic friction performance. Materials **17**, 2765 (2024). https://doi.org/10.3390/ma17112765
8. N. Kladovasilakis, S. Pemas, E.M. Pechlivani, Computer-aided design of 3D-printed clay-based composite mortars reinforced with bioinspired lattice structures. Biomimetics **9**, 424 (2024). https://doi.org/10.3390/biomimetics9070424

9. S. Kim, X. Tang, G.R. Chehab, Laboratory study of geogrid reinforcement in Portland cement concrete. In *Pavement Cracking: Mechanisms, Modeling, Detection, Testing and Case Histories*. (Routledge and CRC Press, Taylor & Francis Group, Oxfordshire, England, UK 2008) pp. 769–778. https://doi.org/10.1201/9780203882191.ch75

10. A. Efstathiadis, I. Symeonidou, K. Tsongas, E.K. Tzimtzimis, D. Tzetzis, 3D printed voronoi structures inspired by paracentrotus lividus shells. Designs **7**, 113 (2023). https://doi.org/10.3390/designs7050113

11. N. Kladovasilakis, P. Charalampous, A. Boumpakis, T. Kontodina, K. Tsongas, D. Tzetzis, I. Kostavelis, P. Givissis, D. Tzovaras, Development of biodegradable customized tibial scaffold with advanced architected materials utilizing additive manufacturing. J. Mech. Behav. Biomed. Mater. **141**, 105796 (2023). https://doi.org/10.1016/j.jmbbm.2023.105796

12. M. Rezapourian, N. Kamboj, I. Jasiuk, I. Hussainova, Biomimetic design of implants for long bone critical-sized defects. J. Mech. Behav. Biomed. Mater. **134**, 105370 (2022). https://doi.org/10.1016/j.jmbbm.2022.105370

13. G. Bergmann, F. Graichen, A. Rohlmann, Hip joint loading during walking and Running. Measured Two Patients. J. Biomech. **26**(8), 969–990 (1993). https://doi.org/10.1016/0021-9290(93)90058-M

14. B.K. Madeti, C.S. Rao, B.S.K.S.S. Rao, Free body diagram and static fi-nite element analysis of the human Tibia. Int. J. Biomed. Eng. Technol. **18**(3), 290 (2015). https://doi.org/10.1504/IJBET.2015.070578

15. S. Pheasant, C.M. Haslegrave, Bodyspace: *Anthropometry, Ergonomics and the Design of Work*, 3rd ed. (CRC Press: Boca Raton, FL, USA; Taylor & Francis: London, UK, 2018). https://doi.org/10.1201/9781315375212

16. H.W. Henn, Crash tests and the head injury criterion. Teach. Math. Appl.: Int. J. IMA **17**(4), 162–170 (1998). https://doi.org/10.1093/teamat/17.4.162

17. A.B. Nellippallil, P.R. Berthelson, L. Peterson, R.K. Prabhu, Chapter 10—Ro-bust concept exploration of driver's side vehicular impacts for human-centric crashworthiness. In P. Raj, H. Mark (eds) *Multiscale Biomechanical Modeling of the Brain*. (Academic Press 2022) pp. 153–176, https://doi.org/10.1016/B978-0-12-818144-7.00002-5

18. W.C. Hayes, M.S. Erickson, E.D. Power, Forensic injury biomechanics. Annu. Rev. Biomed. Eng. **9**, 55–86 (2007). https://doi.org/10.1146/annurev.bioeng.9.060906.151946

19. Y. Chen, Q. Wang, C. Wang, P. Gong, Y. Shi, Y. Yu, Z. Liu, Topology optimization design and experimental research of a 3D-printed metal aerospace bracket considering fatigue performance. Appl. Sci. **11**(15), 6671 (2021). https://doi.org/10.3390/app11156671

20. C. Li, I.Y. Kim, J. Jeswiet, Conceptual and detailed design of an automotive engine cradle by using topology, shape, and size optimization. Struct. Multidiscip. Optim. **51**(2), 547–564 (2015). https://doi.org/10.1007/s00158-014-1151-6

21. E. Tyflopoulos, M. Lien, M. Steinert, Optimization of brake Calipers using topology optimization for additive manufacturing. Appl. Sci. **11**(4), 1437 (2021). https://doi.org/10.3390/app11041437

22. N. Kladovasilakis, P. Charalampous, K. Tsongas, I. Kostavelis, D. Tzetzis, D. Tzovaras, Experimental and computational investigation of lattice sandwich structures constructed by additive manufacturing technologies. J. Manuf. Mater. Process. **5**(3), 95 (2021). https://doi.org/10.3390/jmmp5030095

23. N. Kladovasilakis, K. Tsongas, D. Tzetzis, Mechanical and FEA-assisted characterization of fused filament fabricated triply periodic minimal surface structures. J. Compos. Sci. **5**(2), 58 (2021). https://doi.org/10.3390/jcs5020058

24. L.J. Gibson, M.F. Ashby, *Cellular Solids* (Cambridge University Press, 1997). https://doi.org/10.1017/CBO9781139878326

25. M.M. Sychov, L.A. Lebedev, S.V. Dyachenko, L.A. Nefedova, Mechanical properties of energy-absorbing structures with triply periodic minimal surface topology. Acta Astronaut. **150**, 81–84 (2018). https://doi.org/10.1016/j.actaastro.2017.12.034

26. S.F. Fischer, Energy absorption efficiency of open-cell pure aluminum foams. Mater. Lett. **184**, 208–210 (2016). https://doi.org/10.1016/j.matlet.2016.08.061

27. X. Li, X. Yu, J.W. Chua, H.P. Lee, J. Ding, W. Zhai, Microlattice metamaterials with simultaneous superior acoustic and mechanical energy absorption. Small **17**(24), 2100336 (2021). https://doi.org/10.1002/smll.202100336

28. G. Ciaburro, G. Iannace, L. Ricciotti, A. Apicella, V. Perrotta, R. Aversa, Acoustic applications of a foamed geopolymeric-architected metamaterial. Appl. Sci. **14**, 1207 (2024). https://doi.org/10.3390/app14031207

29. W. Du, L. Luo, S. Wang, J. Liu, B. Sunden, Heat transfer and flow structure in a detached latticework duct. Appl. Therm. Eng. **155**, 24–39 (2019). https://doi.org/10.1016/j.applthermaleng.2019.03.148

30. B. Li, H.C. Zeng, Architecture and preparation of hollow catalytic devices. Adv. Mater. **31**(38), 1801104 (2019). https://doi.org/10.1002/adma.201801104

31. M.S. Saleh, J. Li, J. Park, R. Panat, 3D Printed hierarchically-porous micro-lattice electrode materials for exceptionally high specific capacity and areal capacity lithium ion batteries. Addit. Manuf. **23**, 70–78 (2018). https://doi.org/10.1016/j.addma.2018.07.006

32. M. Ouda, O. Al-Ketan, N. Sreedhar, M.I. Hasan Ali, R.K. Abu Al-Rub, S. Hong, H.A. Arafat, Novel static mixers based on triply periodic minimal surface (TPMS) architectures. J. Environ. Chem. Eng. **8**(5), 104289 (2020). https://doi.org/10.1016/j.jece.2020.104289

33. S.W. Carlson, D.D. Goetz, S.S. Liu, J.J. Greiner, J.J. Callaghan, Minimum 10-year follow-up of cementless total hip arthroplasty using a contemporary triple-tapered titanium stem. J. Arthroplasty **31**(10), 2231–2236 (2016). https://doi.org/10.1016/j.arth.2016.04.037

34. N. Kladovasilakis, K. Tsongas, P. Kyratsis, D. Tzetzis, Finite Element Analysis of Hip Implants with Additive Manufactured Lattice Internal Geometry. International Journal of Modern Manufacturing Technologies (2020), 12 (3 Special Issue).

Chapter 7
Conclusion and Future Research in Topology Optimization

This book investigated and presented innovations in topology optimization, covering topics from generative design to architected materials and practical case studies. It discussed both the element-based and the discrete/truss-based approaches, including their working principles, implementation methodologies, and real-world applications. The main conclusions highlight that the element-based approach has achieved a high-technology readiness level, leading to commercialization primarily in high-value sectors like aerospace and automotive industries. On the other hand, topology optimization using architected materials has dominated the scientific community due to its superior advantages, although its technology readiness remains relatively low, with few products in the market. The book aims to consolidate existing knowledge and provide a comprehensive roadmap to facilitate the implementation of topology optimization approaches and support their industrialization. Additionally, several research gaps and opportunities in the field of topology optimization and architected materials have been identified, leading to proposals for future research to address these gaps and align scientific efforts with real-world applications:

- Development of topology optimization software focused on computational efficiency through the integration of artificial intelligence.
- Development of sophisticated design software that enables the design of numerous architected materials and allows the employment of further optimization processes and their integration of a component's design.
- Investigation of the micro-porosity and dimensional accuracy of additively manufactured architected materials utilizing high-fidelity imaging processes, such as CT scans, in order to obtain more accurate and reliable FEMs.
- Development of a material library cataloging examined architected materials by construction type, in order to create a useful engineering tool and facilitate the employment of these materials in real-life applications.
- Integration of machine learning tools to correlate design-related parameters with the physical and mechanical properties of the employed architected materials.

N. Kladovasilakis et al., *Innovations in Topology Optimization*,
SpringerBriefs in Applied Sciences and Technology,
https://doi.org/10.1007/978-3-031-77700-4_7

- Assessment of the crashworthiness of the developed hybrid architected materials utilizing dynamic tests.
- Examination of the fatigue properties of the examined and developed architected materials.
- Optimization of the fabrication process for the IPCs with architected materials as core structures.
- Evaluation of the diffusion and distribution of the second phase in IPC structures in order to achieve optimal structures.
- Development and examination of architected materials fabricated with multiple construction materials utilizing multi-material AM technologies for both metallic and polymer feedstock materials.
- Utilization of composites with nanoparticles as construction materials for architected materials in order to enhance the physical and mechanical properties of the final structures.
- Implementation of additional optimization processes on the auxetic architected materials to improve performance and facilitate their utilization in 4D printing applications.